Bianca Lim

Boron-oxygen-related defects in crystalline silicon

Bianca Lim

Boron-oxygen-related defects in crystalline silicon

Formation, recovery kinetics, and impact of compensation

Südwestdeutscher Verlag für Hochschulschriften

Impressum/Imprint (nur für Deutschland/only for Germany)
Bibliografische Information der Deutschen Nationalbibliothek: Die Deutsche Nationalbibliothek verzeichnet diese Publikation in der Deutschen Nationalbibliografie; detaillierte bibliografische Daten sind im Internet über http://dnb.d-nb.de abrufbar.
Alle in diesem Buch genannten Marken und Produktnamen unterliegen warenzeichen-, marken- oder patentrechtlichem Schutz bzw. sind Warenzeichen oder eingetragene Warenzeichen der jeweiligen Inhaber. Die Wiedergabe von Marken, Produktnamen, Gebrauchsnamen, Handelsnamen, Warenbezeichnungen u.s.w. in diesem Werk berechtigt auch ohne besondere Kennzeichnung nicht zu der Annahme, dass solche Namen im Sinne der Warenzeichen- und Markenschutzgesetzgebung als frei zu betrachten wären und daher von jedermann benutzt werden dürften.

Coverbild: www.ingimage.com

Verlag: Südwestdeutscher Verlag für Hochschulschriften GmbH & Co. KG
Heinrich-Böcking-Str. 6-8, 66121 Saarbrücken, Deutschland
Telefon +49 681 37 20 271-1, Telefax +49 681 37 20 271-0
Email: info@svh-verlag.de

Zugl.: Hannover, LUH, Diss., 2011

Herstellung in Deutschland (siehe letzte Seite)
ISBN: 978-3-8381-3307-2

Imprint (only for USA, GB)
Bibliographic information published by the Deutsche Nationalbibliothek: The Deutsche Nationalbibliothek lists this publication in the Deutsche Nationalbibliografie; detailed bibliographic data are available in the Internet at http://dnb.d-nb.de.
Any brand names and product names mentioned in this book are subject to trademark, brand or patent protection and are trademarks or registered trademarks of their respective holders. The use of brand names, product names, common names, trade names, product descriptions etc. even without a particular marking in this works is in no way to be construed to mean that such names may be regarded as unrestricted in respect of trademark and brand protection legislation and could thus be used by anyone.

Cover image: www.ingimage.com

Publisher: Südwestdeutscher Verlag für Hochschulschriften GmbH & Co. KG
Heinrich-Böcking-Str. 6-8, 66121 Saarbrücken, Germany
Phone +49 681 37 20 271-1, Fax +49 681 37 20 271-0
Email: info@svh-verlag.de

Printed in the U.S.A.
Printed in the U.K. by (see last page)
ISBN: 978-3-8381-3307-2

Copyright © 2012 by the author and Südwestdeutscher Verlag für Hochschulschriften GmbH & Co. KG and licensors
All rights reserved. Saarbrücken 2012

Abstract

This work investigates the impact of boron-oxygen-related recombination centers on the carrier lifetime and on solar cell parameters in crystalline silicon. Degradation of the carrier lifetime in boron-doped oxygen-rich crystalline silicon under illumination at room temperature has been known for a long time and has been intensely studied during the last fifteen years. In these studies, the effective defect concentration was found to depend linearly on the substitutional boron concentration and quadratically on the interstitial oxygen concentration. In addition, the defect generation rate constant was found to increase quadratically with the square of the boron concentration. Based on these findings, a defect model was developed, in which the recombination-active defect consists of one substitutional boron atom and an interstitial oxygen dimer. However, past studies mainly investigated silicon which was exclusively doped with boron. As a result, no distinction between the substitutional boron concentration and the hole concentration was made. Using compensated silicon doped with both boron and phosphorus, however, the substitutional boron concentration and the hole concentration can be investigated separately. This was done in this work. The free carrier concentration in compensated silicon equals the difference between boron and phosphorus concentrations. If the material contains more boron than phosphorus, it will have p-type conductivity. If the material contains more phosphorus than boron, it will have n-type conductivity. Note that light-induced degradation is observed in both cases. Investigating light-induced degradation in compensated p-type silicon, it is shown that the effective defect concentration actually depends on the hole concentration and not on the boron concentration. In addition, the defect generation rate constant is found to depend on the square of the hole concentration. These results cannot be explained with the established defect model. Therefore, an alternative defect model is discussed and experimentally verified. It has also been known for a long time that the boron-oxygen-related defect can be annihilated through short annealing in darkness, resulting in a recovery of the carrier lifetime. However, this annihilated state is not stable and renewed illumination thus results in renewed degradation. In this work, this annihilation step is investigated for the first time in compensated silicon. These experiments reveal an inverse dependence of the annihilation rate constant on the free carrier concentration. In addition, defect annihilation is found to take considerably longer in compensated n-type silicon than in p-type silicon. Apart from understanding the underlying defect mechanism, a way to permanently reduce the defect concentration or even completely avoid its generation is a matter of great interest, especially with regard to increasing the energy conversion efficiency of solar cells fabricated on boron-doped oxygen-rich silicon. The second part of this work thus examines a procedure, which is capable of permanently decreasing the defect concentration: illumination at temperatures between 135°C and 220°C. In order to identify the mechanisms behind this deactivation, the deactivation rate constant is examined in this work for the first time as a function of boron concentration, net doping concentration and interstitial oxygen concentration, respectively. An inverse dependence of the deactivation rate constant on the boron concentration as well as an inverse quadratic dependence on the interstitial oxygen concentration is found. In addition, processing steps at high temperature result in an increase of the deactivation rate constant by a factor of 4, whereas deposition of a silicon nitride layer using plasma-enhanced chemical vapor deposition results in an increase of the deactivation rate constant by a factor of 5. Permanent deactivation is not observed in compensated n-type silicon. While the lifetime recovers under illumination at elevated temperature, this state is not stable under illumination at room temperature for compensated n-type silicon. Finally, the deactivation treatment is successfully applied to silicon solar cells, by which for the first time a stable efficiency above 20% is obtained for a solar cell fabricated on low-resistivity B-doped Czochralski-grown silicon.

Contents

1 Introduction   **5**

2 Characterization techniques   **9**
    2.1 Carrier lifetime measurements 9
        2.1.1 Quasi-steady-state photoconductance (QSSPC) 9
        2.1.2 Microwave-detected photoconductance decay (MW-PCD) 15
    2.2 Electrochemical capacitance voltage technique 17

3 Compensation in mono- and multicrystalline silicon   **19**
    3.1 The Czochralski-growth process 20
    3.2 Block-casting of silicon 21
    3.3 Dopant concentrations in compensated silicon ingots 22
    3.4 Majority- and minority-carrier mobilities in compensated crystalline silicon . . . 23
        3.4.1 Block-cast multicrystalline silicon 23
        3.4.2 Monocrystalline Czochralski-grown silicon 27
    3.5 Chapter summary 30

4 Generation of boron-oxygen-related recombination centers   **37**
    4.1 Review of previous experimental and theoretical work 38
    4.2 Light-induced degradation in dopant-compensated Cz-Si 41
        4.2.1 Compensated p-type Cz-Si 42
        4.2.2 Compensated n-type Cz-Si 45
    4.3 Chapter summary 47

5 Annihilation of boron-oxygen-related recombination centers   **49**
    5.1 Exclusively boron-doped p-type Cz-Si 49
    5.2 Boron- and phosphorus-doped Cz-Si 51
        5.2.1 Boron- and phosphorus-doped p-type Cz-Si 51
        5.2.2 Boron- and phosphorus-doped n-type Cz-Si 54
    5.3 Chapter summary 56

6 Permanent deactivation of boron-oxygen-related recombination centers — 57

- 6.1 Deactivation of the BO complex in p-type Cz-Si 59
  - 6.1.1 Impact of phosphorus diffusion 59
  - 6.1.2 Impact of silicon nitride deposition 62
  - 6.1.3 Impact of boron concentration 65
  - 6.1.4 Impact of interstitial oxygen concentration 68
  - 6.1.5 Impact of thermal donors 70
  - 6.1.6 Impact of compensation 72
  - 6.1.7 Potential of the carrier lifetime after permanent deactivation of the BO defect 74
- 6.2 Stability of the deactivated state 75
  - 6.2.1 Partial degradation under illumination at room temperature 75
  - 6.2.2 Complete degradation during long-term illumination at elevated temperature 77
  - 6.2.3 Complete degradation through extended annealing in darkness 78
- 6.3 Boron- and phosphorus-doped n-type Cz-Si 79
- 6.4 Chapter summary 81

7 Defect models — 83

- 7.1 The B_sO_{2i}-model 83
- 7.2 The B_iO_{2i}-model 85
- 7.3 Chapter summary 88

8 Application of the deactivation procedure to solar cells — 91

- 8.1 Screen-printed solar cells 91
  - 8.1.1 Exclusively boron-doped Cz-Si 92
  - 8.1.2 Boron- and phosphorus-doped Cz-Si 93
- 8.2 Efficiency potential of screen-printed solar cells according to one dimensional solar cell simulations 96
- 8.3 High-efficiency RISE-EWT solar cells 101
- 8.4 Chapter summary 103

9 Summary — 105

References — 109

List of publications — 119

1 Introduction

Crystalline silicon solar cells constitute the bigger part (~85% in 2010 [1]) in photovoltaic production nowadays. Compared to low-cost, low-efficiency thin-film technology and high-cost, high-efficiency multi-junction solar cells, they offer a compromise of cost and obtainable energy conversion efficiency.

The cost of a solar cell can be split into two factors: (1) the material and (2) the production processes. Naturally, both factors are also important for a solar cell's efficiency. A good and easily obtainable measure of material quality is the (excess) carrier lifetime, i.e. the time between generation of excess carriers and their recombination, which is studied in this work.

The carrier lifetime increases with decreasing overall recombination. In silicon, recombination mainly takes place via defects, such as impurities or crystallographic defects. Decreasing impurity concentrations and increasing crystal quality thus improve the lifetime. Of course, extended purification and enhanced crystallization also increase material cost. It should be noted, however, that at some point, depending on the used solar cell structure and the production processes, other aspects than material quality (and thus lifetime) will limit the solar cell efficiency. The challenge is thus to balance the cost of purification and crystallization against the benefit for cell efficiency.

The vast majority of crystalline silicon used in current production is boron-doped mono- or multicrystalline silicon. Monocrystalline silicon can further be split into two groups: Float zone silicon (FZ-Si) and Czochralski-grown silicon (Cz-Si). Monocrystalline silicon has better crystal quality and in general less impurities than multicrystalline silicon (mc-Si) but the fabrication of monocrystalline ingots is more expensive than that of multicrystalline ingots. FZ-Si in particular is too expensive to be used in solar cell production. Cz-Si on the other hand can be produced cost-effectively. However, Cz-Si contains large amounts of oxygen, which is particularly harmful in combination with boron-doping.

It has been known for a long time that the carrier lifetime in boron-doped oxygen-rich silicon degrades significantly under illumination at room temperature [2]. As a result, the efficiency of solar cells fabricated on such material degrades by up to 10% relative [3]. This so-called light-induced degradation (LID) has been intensely studied [2–19] and the effect is firmly linked to the simultaneous presence of both boron and oxygen [4]. In particular, higher quantities of either boron or oxygen result in lower lifetimes after degradation.

In current solar cell production, this issue is addressed by using low doping concentrations of around $N_A = 5 \times 10^{15}$ cm^{-3}, which balances the gain in degraded lifetime against the losses due to increased resistivity of the material. However, in order to further reduce material cost, new purification routes are explored, which result in much higher boron concentrations in the material. It is thus essential to examine how these increased boron concentrations affect the degraded lifetime in the material.

The first part of this work examines the boron-oxygen-related recombination center in silicon doped with both boron and phosphorus. In such compensated silicon, the net doping concentration differs from the boron concentration, which allows new insights regarding the defect composition and its generation kinetics. In the second part of this work, we examine a recently proposed deactivation procedure [20, 21], which promises to permanently reduce the defect concentration, thus leading to improved lifetimes that are stable under illumination at room temperature.

Chapter 2 gives a short overview of the characterization techniques and measurement setups used to measure carrier lifetime and doping concentration.

Chapter 3 briefly outlines the value chain of silicon wafers used in photovoltaic production today. Subsequently, the origin of compensation in material made from less refined feedstock and its effect on the carrier mobility are presented. A significant reduction of the mobility is observed in the studied material, which needs to be considered in the analysis of the lifetime measurements.

In Chapter 4, the generation of boron-oxygen-related recombination centers under illumination (at room temperature) in compensated p- and n-type Cz-Si is investigated. Through this, the impact of the doping concentration and the boron concentration on the generation of boron-oxygen-related defects can be investigated separately.

Chapter 5 examines the defect annihilation during annealing in darkness in compensated p- and n-type Cz-Si. Even though this defect annihilation does not result in stable lifetimes, it is an important aspect of the boron-oxygen-related defect and can contribute useful information on the defect transformation mechanism.

In Chapter 6, a means to permanently deactivate the boron-oxygen defect is investigated. The impact of various processing steps and material characteristics is studied in B-doped p-type Cz-Si and a large number of parameters is found to influence the deactivation process. In addition, the procedure is applied to compensated n-type Cz-Si, where it is found to be ineffective.

Chapter 7 discusses two defect models for the boron-oxygen-related recombination center. The standard model, in which the defect is composed of one substitutional boron atom and an interstitial oxygen dimer, is found to be incapable of explaining the new experimental data obtained on compensated p-type silicon. On the other hand, an alternative model, in which the defect is composed of one interstitial boron atom and an interstitial oxygen dimer, is found to explain all experimental results presented in this work.

In Chapter 8, the permanent deactivation treatment is applied to solar cells. An increase in solar cell efficiency by 1.2% absolute is observed in industrial screen-printed solar cells, while a stable efficiency of 20.4% is obtained on a high-efficiency RISE-EWT solar cell fabricated on low-resistivity B-doped p-type Cz-Si. Using the lifetime data from Chapter 6, one-dimensional solar cell simulations are performed in order to investigate the potential of screen-printed solar cells fabricated on low-resistivity B-doped Cz-Si. By using an advanced cell concept, which includes passivation of the emitter and rear, an efficiency potential of 20.6% is found for solar cells fabricated on 0.8 Ω cm B-doped p-type Cz-Si after permanent deactivation of the BO defect.

Finally, Chapter 9 summarizes the results presented in this work.

2 Characterization techniques

2.1 Carrier lifetime measurements

2.1.1 Quasi-steady-state photoconductance (QSSPC)

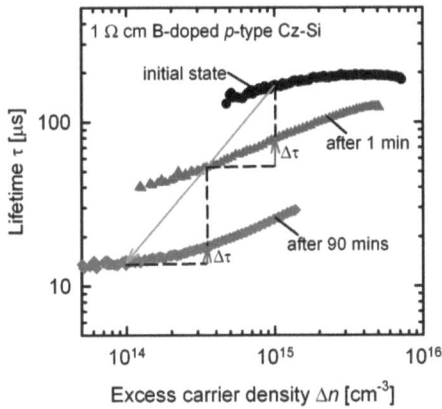

Figure 2.1: Example of the injection-dependent carrier lifetime measured using the quasi-steady-state photoconductance technique. The red arrows indicate the error made by determining a single lifetime value at a fixed light intensity.

The quasi-steady-state photoconductance (QSSPC) technique [22] offers a fast and accurate means to measure the effective carrier lifetime τ_{eff} over a wide range of excess carrier densities Δn. This in turn offers the possibility to extract the lifetime at a fixed injection level, which is important when comparing the lifetimes of different samples or when monitoring the change of lifetime in a single sample over the course of time. An example of the injection-dependent carrier lifetime measured via QSSPC is shown in Fig. 2.1.

The lifetime in a 1 Ω cm B-doped Czochralski-grown silicon sample is measured at different stages of lifetime degradation. The amount of excess carriers at a given generation rate G is proportional to the lifetime τ (as is demonstrated by the decreasing maximum excess carrier density Δn_{max} with decreasing carrier lifetime τ). The orange arrow indicates this shift in excess

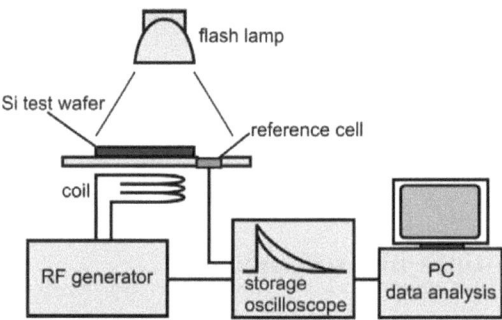

Figure 2.2: Schematic of the setup used for quasi-steady-state photoconductance measurements [23].

carrier density at a fixed generation rate G. The red arrows mark the difference in measured lifetime at this new excess carrier density and the previously analyzed Δn. Obviously, this error increases with increasing difference of the lifetimes as well as with increasing injection dependence of the lifetime.

In this work, a Sinton Instruments WCT-100 as well as a WCT-120 lifetime tester are used for QSSPC measurements. A schematic of the measurement setup is depicted in Fig. 2.2. The silicon test wafer is inductively coupled to a coil, which in turn is part of an rf-bridge circuit. The conductance σ of the silicon wafer is a quadratic function of the output voltage V_{wafer} of the rf-bridge circuit. Using a set of reference wafers with known conductance, a calibration curve of the form $\sigma = a\,V_{\text{wafer}}^2 + b\,V_{\text{wafer}} + c$ is determined, as shown in Fig. 2.3. This curve can then be used to determine the conductance of any silicon sample. Note that the coil has a diameter of 18 mm and accordingly covers an area of approximately 250 mm^2. All measurements are thus averaged over this area.

For a lifetime measurement, the silicon test wafer is illuminated by a flash (decay time approximately 2.1 ms), which generates excess carriers in the sample. The original spectrum of the flash is quite similar to the solar spectrum, however, in order to obtain a homogeneous photogeneration throughout the sample, the light of the flash is filtered by a 700 nm IR-pass filter. The intensity of the flash is monitored as a function of time via a calibrated solar cell.

The short-circuit current I_{SC} of the reference cell is known to depend linearly on the light intensity I, thus $I = c\,I_{\text{SC}}$. The calibration factor c is obtained by measuring I_{SC} under standard testing conditions (25°C, AM1.5G spectrum, 100 mW/cm^2 light intensity). In order to reduce the signal-to-noise ratio, the current is also converted to a voltage V_{light} via a resistor. At the same time, the generated excess carriers increase the conductance of the sample (photoconductance). This increase is measured by the rf-bridge circuit. Using an oscilloscope, both the photoconductance and the light intensity are monitored as a function of time t.

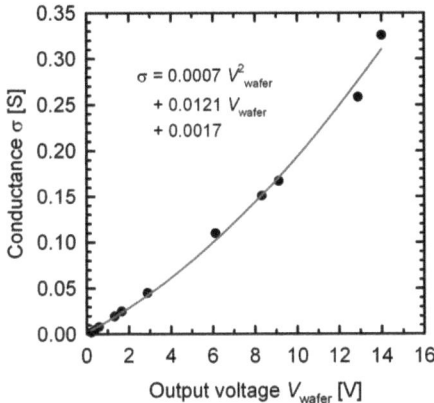

Figure 2.3: Calibration function of the rf-bridge circuit.

In general, the conductance σ of a silicon wafer is a product of carrier density and carrier mobility:

$$\sigma = q \int_0^W (n \, \mu_n + p \, \mu_p) \, dz. \tag{2.1}$$

Here, q is the elementary charge, W is the sample thickness, n and p are the electron and hole concentrations, respectively, and μ_n and μ_p are the electron and hole mobilities, respectively. Note that the carrier concentrations in this expression are a function of position on the z-axis, while the mobilities are a function of the carrier concentration at position z.

During a QSSPC measurement, the quantity of interest is the excess carrier density Δn, and hence the photoconductance. It is thus convenient to split the measured total conductance σ of the sample into the base conductance σ_{base} (which is then treated as an offset) and the photoconductance $\Delta\sigma$:

$$\sigma = \sigma_{\text{base}} + \Delta\sigma. \tag{2.2}$$

Since the photogeneration of excess carriers is quite homogeneous, Eq. 2.1 can be simplified by introducing an average excess carrier density Δn_{av} (if the recombination at the surface is small). The photoconductance $\Delta\sigma$ is then given by:

$$\Delta\sigma \approx q \, W \, (\mu_n + \mu_p) \, \Delta n_{\text{av}}. \tag{2.3}$$

In order to transform the measured photoconductance into the (average) excess carrier density Δn in the sample, one then only needs to know the sum of electron mobility μ_n and hole mobility μ_p. In exclusively boron-doped p-type silicon and in exclusively phosphorus-doped

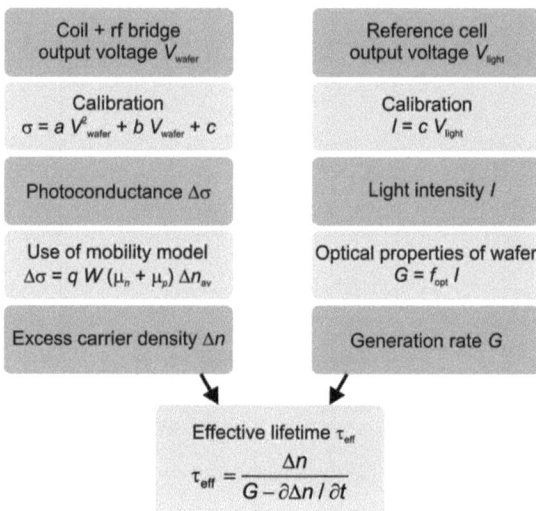

Figure 2.4: Flow chart of the data analysis during quasi-steady-state photoconductance measurements.

n-type silicon, these mobilities are well known and can be described by the semi-empirical expression [24]:

$$(\mu_n + \mu_p) = 1800 \frac{1 + \exp\left[0.8431 \ln\left(\frac{\Delta n + N_{\text{dop}}}{N_{\text{ref}}}\right)\right]}{1 + 8.36 \exp\left[0.8431 \ln\left(\frac{\Delta n + N_{\text{dop}}}{N_{\text{ref}}}\right)\right]}, \quad (2.4)$$

where N_{dop} is the doping concentration and $N_{\text{ref}} = 1.2 \times 10^{18}$ cm^{-3}.

In compensated silicon (doped with both boron and phosphorus, or B-doped and compensated by thermal donors), however, a considerable reduction of carrier mobilities compared to mobilities in non-compensated silicon has been observed [25–27]. A similar reduction is observed in the dopant-compensated samples investigated in this work (see Sec. 3.4.2). Consequently, the prefactor in Eq. 2.4 is adjusted to measured values of μ_n and μ_p for the analysis of the lifetime data obtained on compensated Cz-Si. Note that this correction is only valid for low injection conditions, however, these are of interest in this work.

The calibrated reference cell is used to obtain the light intensity at time t. Depending on the optical parameters of the sample f_{opt}, such as anti-reflection coating and texturization, the generation rate G of excess carriers can be calculated.

Knowing both the excess carrier density Δn and the generation rate G allows the determination of the carrier lifetime τ. The time dependence of the excess carrier density Δn is given by the continuity equation:

$$\frac{\partial \Delta n}{\partial t} = G(t) - U(t) + \frac{1}{q}\nabla J, \tag{2.5}$$

where U is the recombination rate and J is the current density. Since the photogeneration is very homogeneous and the surfaces are well passivated, the carrier density in the sample can be assumed to be spatially uniform. As a result, the last term in Eq. 2.5 can be neglected. The recombination rate U can further be written as $U = \Delta n/\tau_{\text{eff}}$, which yields for the effective lifetime τ_{eff} [28]:

$$\tau_{\text{eff}} = \frac{\Delta n}{G - \partial \Delta n/\partial t}. \tag{2.6}$$

A flow chart of the analysis is shown in Fig. 2.4: on the left hand side, the inductive coupling of the wafer to the coil and the rf-bridge circuit results in a change of the output voltage V_{wafer}. Using the calibration function shown in Fig. 2.3, the photoconductance $\Delta\sigma$ is obtained. In combination with a semi-empirical mobility model (Eq. 2.4) and measured mobility data (see Section 3.4), the conductance is then converted into the excess carrier density Δn. On the right-hand side of the flow chart, the light intensity of the flash is detected by the reference cell. Taking the optical properties of the sample f_{opt} into account, the generation rate G is obtained. Inserting Δn and G into Eq. 2.6 then yields the effective lifetime τ_{eff}.

Measurement uncertainty

The measurement uncertainty of the Sinton Instruments WCT-100 lifetime tester has been analyzed in detail by Berge in 1998 [29], who investigated

- heating of the wafer due to high-intensity illumination,
- heating of the wafer due to energy dissipation from the rf-bridge,
- impact of the spectrum of the flash lamp,
- impact of the optical properties of the sample,
- low signal-to-noise ratio,
- digitalization, and
- position of the flash lamp.

In summary, Berge found that the major source of uncertainty was the possible difference between the photogeneration rate G in the reference solar cell and in the sample, i.e., the factor accounting for the optical parameters of the sample f_{opt}. Low signal-to-noise ratios and errors resulting from digitalization, on the other hand, become significant at low excess carrier densities $\Delta n < 10^{14}$ cm^{-3}.

Recently, the measurement uncertainty of photoconductance lifetime measurements via inductive coupling was also investigated by McIntosh and Sinton [30]. Their list of sources of uncertainty includes

- the calibration of the rf-bridge circuit,
- the calibration of the reference solar cell,
- the width of the quasi-neutral base W,
- the output voltages V_wafer and V_light,
- the sum of carrier mobilities, and
- the base doping concentration.

Taking all aspects into account, they estimated an uncertainty of $\pm 10.9\%$ for measurements done under quasi-steady-state conditions and an uncertainty of $\pm 8.6\%$ for PCD measurements. However, they stressed that the impact of each item on the list strongly depends on the actual measurement setup and the care that was taken to determine the values of interest.

It should also be mentioned that the Sinton Instruments WCT lifetime testers have an *absolute* uncertainty (caused by systematic errors) as well as a *relative* uncertainty (caused by statistical errors). The absolute uncertainty is important when investigating absolute lifetime values, e.g. when comparing different lifetime measurement techniques. However, when the time dependence of the carrier lifetime under certain circumstances is investigated, the relative uncertainty is of more importance. In the course of this work, this relative uncertainty was found to be less than 5%.

Photoconductance decay (PCD) measurements

The Sinton Instruments lifetime testers can also be used for photoconductance decay measurements. For this, a much shorter flash (decay time approximately 30 μs) is used. As a result, the generation rate G vanishes and Eq. 2.6 simplifies to:

$$\tau_\text{eff} = -\frac{\Delta n}{\partial \Delta n / \partial t}. \qquad (2.7)$$

Note that since the optical parameters of the sample are now irrelevant the (absolute) uncertainty of the measurement technique is notably reduced, as was also pointed out by McIntosh and Sinton [30]. In addition, a wide range of excess carrier densities Δn is covered during the decay of the photoconductance and the effective lifetime is thus again measured as a function of Δn.

In order to use Eq. 2.7, the effective lifetime τ_eff of the sample needs to be much larger than the decay time constant of the flash. For the present setup, this condition is fulfilled when $\tau_\text{eff} > 200$ μs.

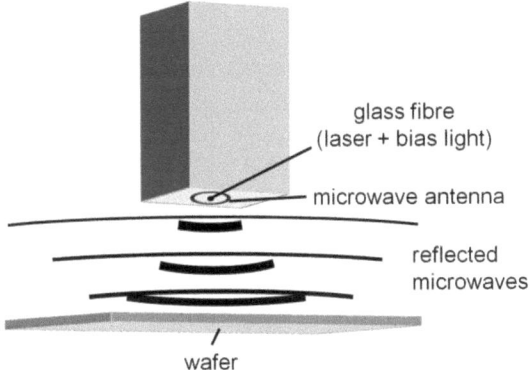

Figure 2.5: Schematic of the setup used for microwave-detected photoconductance decay measurements.

2.1.2 Microwave-detected photoconductance decay (MW-PCD)

For spatially resolved carrier lifetime measurements, microwave-detected photoconductance decay (MW-PCD) measurements were performed using a SEMILAB WT-2000 tool. A simplified schematic of an MW-PCD setup is depicted in Fig. 2.5. The silicon test wafer is placed under a microwave source and excess carriers are generated by a short laser pulse (wavelength 908 nm, pulse length 200 ns). Since the microwaves are reflected by free carriers in the silicon, the amplitude of the reflected microwave signal is a measure of the carrier density Δn in the sample. By monitoring the microwave amplitude as a function of time, the photoconductance decay can thus be monitored.

Note, however, that the reflected microwave power P is in general a non-linear function of the conductance σ (and accordingly of Δn), i.e., only in the small signal case $\Delta \sigma \ll \sigma$ is $\Delta \sigma$ proportional to ΔP. Since a small $\Delta \sigma$ corresponds to a small Δn, additional excess carriers are generated by a white bias light.

As already mentioned above, the time dependence of the excess carrier density Δn is described by the continuity equation. However, in contrast to the QSSPC measurements, the generation of excess carriers during MW-PCD measurements is non-uniform. As a consequence, the carrier distribution in the sample is inhomogeneous and the divergence of the current density ∇J is no longer negligible. As a result, the general form of the continuity equation needs to be examined [31]:

$$\frac{\partial \Delta n(x,t)}{\partial t} = G - \frac{\Delta n}{\tau_b} + D_a \frac{\partial^2 \Delta n(x,t)}{\partial x^2} - \mu_a E \frac{\partial \Delta n(x,t)}{\partial x}. \tag{2.8}$$

Here, the third term describes the diffusion current and the fourth term describes the drift current of charge carriers. τ_b is the carrier lifetime in the bulk, $D_a = (n+p)/(n/D_p + p/D_n)$

is the ambipolar diffusion coefficient and $\mu_a = (n-p)/(n/\mu_p + p/\mu_n)$ is the ambipolar mobility.

In the low-injection case $\Delta n \ll p_0$, the electric field E is negligible and the continuity equation simplifies to the diffusion equation. In addition, the ambipolar diffusion coefficient D_a can be replaced by the minority-carrier diffusion coefficient (D_n in p-type silicon and D_p in n-type silicon):

$$\frac{\partial \Delta n(x,t)}{\partial t} = G - \frac{\Delta n}{\tau_b} + D_n \frac{\partial^2 \Delta n(x,t)}{\partial x^2}. \tag{2.9}$$

In the case of equally passivated surfaces, two boundary conditions apply:

$$\left.\frac{\partial \Delta n(x,t)}{\partial x}\right|_{x=0} = S\frac{\Delta n(0,t)}{D_n} \quad \text{and} \quad \left.\frac{\partial \Delta n(x,t)}{\partial x}\right|_{x=W} = -S\frac{\Delta n(W,t)}{D_n}, \tag{2.10}$$

where W is the sample thickness and S is the surface recombination velocity. Under a pulsed excitation, the solution to Eq. 2.9 can be written as

$$\Delta n(x,t) = \sum_{m=0}^{\infty} A_m \exp\left(-\frac{t}{\tau_m}\right), \tag{2.11}$$

where the coefficients A_m depend on the sample thickness, the absorption coefficient $\alpha(\lambda)$ for the wavelength of optical excitation (i.e., in the present case 908 nm) and the surface recombination velocity S. The decay time constants τ_m are given by

$$\frac{1}{\tau_m} = \frac{1}{\tau_b} + \alpha_m^2 D_n. \tag{2.12}$$

Here, α_m are given by the transcendental equation

$$\tan \alpha_m W = \frac{2D_n \alpha_m S}{D_n^2 \alpha_m^2 - S^2}, \tag{2.13}$$

which can be solved graphically or numerically. Note that the value of α_m increases monotonously, which means that the decay time constants τ_m decrease with increasing m. Since the fast decay modes fade after a short time, the decay of Δn quickly becomes monoexponential:

$$\Delta n(t) = \Delta n_0 \exp\left(-\frac{t}{\tau_{\text{eff}}}\right). \tag{2.14}$$

Infinite surface recombination velocity

For large values of the surface recombination velocity S, the solution for α_m approaches

$$\alpha_m = \frac{(2m-1)\pi}{W}, \tag{2.15}$$

which results in a decay time constant for the principal mode $m=1$ of

$$\frac{1}{\tau_{\text{eff}}} = \frac{1}{\tau_b} + \left(\frac{\pi}{W}\right)^2 D_{\min}. \tag{2.16}$$

Here, τ_b is the bulk carrier lifetime, W is the wafer thickness, and $D_{\min}$ is the minority-carrier diffusion coefficient. Since $D_{\min}$ relates to the minority-carrier mobility $\mu_{\min}$ via

$$\mu_{\min} = \frac{q}{k_B T} D_{\min}, \qquad (2.17)$$

where k_B is Boltzmann's constant and T is the temperature, the minority-carrier mobility $\mu_{\min}$ can be determined from effective carrier lifetime measurements in the case of very large surface recombination velocities [32].

2.2 Electrochemical capacitance voltage technique

In order to determine the doping concentration N_{dop} in compensated silicon, electrochemical capacitance voltage (ECV) measurements [33] were performed using a WEP CVP21 profiler. The operation mode of the measurement technique is as follows: the silicon wafer is contacted by an electrolyte. As a result, a Schottky barrier is formed through the depletion of carriers at the silicon surface. By applying an external voltage, the width of the depletion region can be changed. In reverse bias, the width increases and accordingly the capacitance of the depletion region decreases, while in forward bias, the width of the depletion region decreases while the capacitance increases.

Determination of the doping concentration is done in reverse bias. The voltage is varied by an ac source, which results in variation of the capacitance C and accordingly in a flow of charge. Knowing the applied voltage V and the capacitance C, the doping concentration N_{dop} can be derived from the Mott-Schottky equation:

$$\frac{1}{C^2} = \frac{-2}{q \, \epsilon_0 \, \epsilon_{\text{Si}} \, A^2 \, N_{\text{dop}}} (V - V_{fb}). \qquad (2.18)$$

Here, q is the elementary charge, ϵ_0 is the vacuum permittivity and ϵ_{Si} is the relative permittivity of silicon. A is the area of the Schottky contact and V_{fb} is the flatband voltage at which the depletion region width is zero. Transposing to N_{dop} then yields

$$N_{\text{dop}} = \frac{-2}{q \, \epsilon_0 \, \epsilon_{\text{Si}} \, A^2 \, \frac{d(1/C^2)}{dV}}. \qquad (2.19)$$

Since the WEP CVP21 setup is designed to measure doping *profiles*, the system is also capable of etching silicon. For this, a 0.1M solution of NH_4F is used. Measurement of a doping profile then consists of cycling between measurement of the doping concentration and etching of a defined amount of silicon.

Note that the doping concentration N_{dop} depends quadratically on the area A of the Schottky contact, which is thus a major source of measurement uncertainty. In the present case, A is defined by a plastic sealing ring, the size of which is known. However, it was recently demonstrated that an external determination of the contact area A, e.g. through inspection of the etching crater with an optical microscope, should always be performed [34]. The thus determined area $A = 0.1142$ cm^2 was subsequently used for the analysis of the measurement data. Due to the low doping concentrations investigated in this work, the ECV measurements nonetheless have an uncertainty of $\pm 10\%$.

3 Compensation in mono- and multicrystalline silicon

Silicon exists in abundance. Unfortunately, it is not found as a pure element but instead exists as silicon dioxide or silicates. In addition, the silicon dioxide and the silicates contain a large variety of other elements in large numbers. The fabrication of pure silicon thus requires extensive refining.

The reduction of silicon dioxide is done in an electric arc furnace using coal. This results in so-called metallurgical-grade silicon, which has a purity of at least 98%. For the microelectronic industry, where purities as high as 99.9999999% (9N) are needed, silicon is further purified via the Siemens process. This process includes the transformation of silicon into gaseous phase (trichlorosilane) and then running multiple distillations. Subsequently, the purified gas is decomposed at over 1100°C in the presence of high-purity silicon rods, at which additional silicon is then deposited. Due to its complexity and the very high temperatures involved, this process is naturally very costly.

In order to reduce the cost of silicon purification, refining routes which avoid the transformation into the gas phase are explored. Examples of such metallurgical purification schemes, which result in so-called upgraded metallurgical silicon (UMG-Si), are the usage of high purity quartz and carbon powder (thus reducing the initial concentration of impurities), the addition of reactants to the silicon melt to bind impurities, and multiple directional solidifications, which extract impurities with low segregation coefficients.

However, the extraction of boron and phosphorus by these means is very difficult, as their segregation coefficients are comparatively large (0.8 and 0.35, respectively, see Tab. 3.1). As a result, silicon which is purified without distillation in the gas phase contains high amounts of boron and phosphorus. The net doping concentration, i.e. the difference of acceptor and donor concentrations, which determines the resistivity of the material, can be adjusted by adding additional dopants, however, the initially high concentrations may already be harmful by themselves. As UMG-Si is a relatively novel material, not much research has so far been undertaken to understand this material and in particular the effects of dopant compensation.

This Chapter briefly describes the two most important crystallization techniques of silicon used for today's photovoltaic silicon production. (1) The Czochralski process that results in monocrystalline silicon and (2) block-casting which results in multicrystalline silicon. In the third Section of this Chapter, the doping profiles of boron and phosphorus in silicon ingots are discussed. Finally, in the last Section, the carrier mobilities in mono- and multicrystalline silicon doped with both boron and phosphorus are investigated.

3.1 The Czochralski-growth process

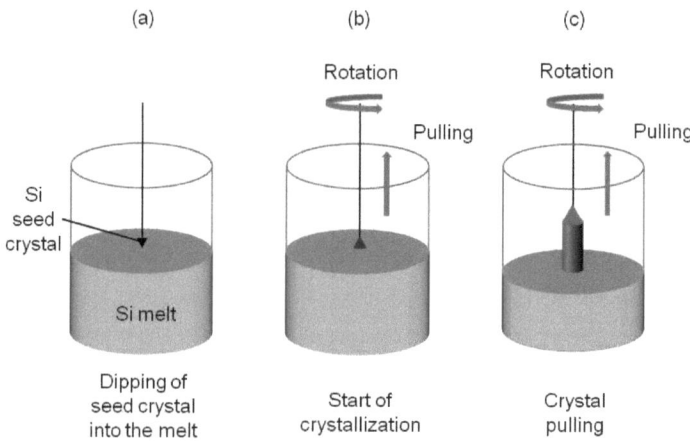

Figure 3.1: Schematic of the Czochralski-growth technique. (a) Dipping of the seed crystal into the silicon melt. (b) Beginning of crystallization with simultaneous rotation and pulling. (c) Pulling of the ingot.

The Czochralski process was developed in 1916, at which time is was used to study the crystallization of metals [35]. However, the principle of crystallization can also be applied to semiconductors or salts. A schematic of the process is shown in Fig. 3.1: (a) At the beginning of the process, the material of interest is molten in a crucible and then kept slightly above the melting point. Subsequently, a seed crystal is submerged into the melt. (b) The melt crystallizes in accordance with the orientation of the seed crystal. (c) Through simultaneous rotation and pulling, the crystal slowly growths in diameter and length.

For the crystallization of silicon, high-purity polycrystalline silicon made via the Siemens process is used as feedstock, while the crucibles, which need to withstand more than 1400°C, are made out of quartz (SiO_2). Dopants, such as boron or phosphorus, are added to the feedstock before melting. By adjusting the temperature, the rotational speed and the pulling speed, the diameter of the ingot can be controlled. In current production, ingots with 200 mm and 300 mm diameter are produced.

Since the melt and the quartz crucible are kept at over 1400°C for a long time, oxygen from the crucible diffuses into the melt and is thus incorporated into the crystal. As a result,

Czochralski-grown silicon contains high amounts of oxygen. Typical values for the interstitial oxygen concentration $[O_i]$ are in the range of 5×10^{17} cm^{-3} to 1×10^{18} cm^{-3}.

3.2 Block-casting of silicon

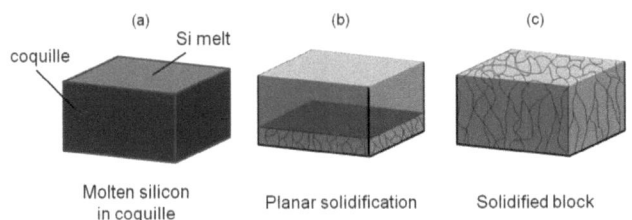

Figure 3.2: Schematic of the block-casting process used to manufacture multicrystalline silicon (mc-Si). (a) The coquille is filled with molten silicon. (b) The silicon solidifies with a planar interface. (c) Completely solidified mc-Si ingot.

A more cost-effective way of crystallization is block-casting. A schematic of the process is depicted in Fig. 3.2: (a) High-purity feedstock is molten in a crucible and subsequently transferred to the so-called coquille. (b) Through controlled cooling of the melt, the silicon slowly solidifies from bottom to top, with a planar solidification interface until (c) the complete block is crystallized.

The coquille is made out of quartz and is additionally coated with silicon nitride (Si_3N_4). This coating acts as a separating agent between the molten silicon and the quartz and facilitates crystallization without the formation of cracks. Block-cast silicon contains considerably less oxygen than Czochralski-grown silicon (typically $[O_i] = (2-3) \times 10^{17}$ cm^{-3}), however, it is known to contain significant amounts of metal impurities [36, 37]. Since high amounts of metal impurities are also observed in block-cast ingots made from high-purity feedstock, it is likely that one source of these impurities is the Si_3N_4 coating.

In contrast to the Czochralski process, crystallization by block-casting does not result in a monocrystalline ingot. Instead, the solidified silicon consists of many crystallites of different size and orientation (so-called grains), which is why it is referred to as multicrystalline silicon (mc-Si). Apart from the many grain-boundaries, mc-Si also contains high amounts of dislocations and in particular dislocation networks. Given such high amounts of metal impurities and crystal defects, it is not surprising that the material quality of mc-Si is in general lower than that of Cz-Si.

3.3 Dopant concentrations in compensated silicon ingots

The solubility of an element in liquid silicon differs from its solubility in solid silicon. As a result, there is a constant redistribution of solutes during crystallization. If the solubility in the liquid phase is higher than that in the solid phase, the concentration of the solute will accumulate in the liquid. As a result, the concentration in the solid increases with increasing amount of solidified portion. This is described by the Scheil equation [38]:

$$C_S = k\, C_0\, (1 - f_S)^{k-1}, \qquad (3.1)$$

where C_S is the concentration in the solid, k is the effective segregation coefficient, C_0 is the initial concentration in the melt and f_S is the fraction of solidified material.

The equilibrium segregation coefficients k_0 of common dopants and metal impurities in silicon are summarized in Tab. 3.1. An example of the boron and phosphorus concentrations in a solidified silicon ingot according to Eq. 3.1 is shown in Fig. 3.3. The concentrations in the melt assumed for the calculation are $[\text{B}]_\text{melt} = 1.2 \times 10^{17}$ cm^{-3} for boron and $[\text{P}]_\text{melt} = 2.0 \times 10^{17}$ cm^{-3} for phosphorus.

Due to its higher segregation coefficient, the concentration of boron (solid blue line) in the part which solidifies first is higher than the concentration of phosphorus (dashed red line), even as $[\text{B}]_\text{melt} < [\text{P}]_\text{melt}$. As a result, the first part of the ingot has p-type conductivity. With advancing solidification, both the boron and phosphorus concentrations increase. However, due to its lower segregation coefficient, the increase of the phosphorus concentration is considerably steeper than the increase of the boron concentration. Consequently, the difference between boron and phosphorus concentrations (dashed green line) decreases with advancing solidification until [B] = [P] (indicated by the vertical black line). After this point, there is more phosphorus than boron in the solidified silicon and the material accordingly has n-type conductivity.

The large variation of net doping concentration, and consequently of the resistivity, over the ingot height is a disadvantage of compensated silicon, since most processes in solar cell production need to be optimized for a specific resistivity. In addition, the portion that has n-type conductivity is discarded, since the vast majority of solar cells is fabricated on p-type silicon.

The position of the transition point depends on the amount of boron and phosphorus in the melt and can be pushed towards later solidification through adding additional boron (or other acceptors). However, by adding boron the total dopant concentration is increased and the overall quality of the material is reduced due to decreasing mobility and carrier lifetimes.

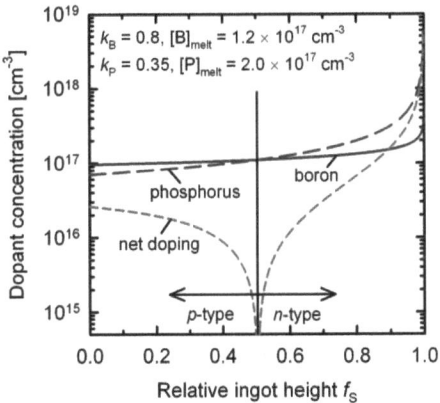

Figure 3.3: Concentration of boron (solid blue line) and phosphorus (dashed red line) in a silicon ingot as a function of relative ingot height f_S as given by the Scheil equation (Eq. 3.1). Also shown is the net doping concentration (dashed green line), i.e. the difference between boron and phosphorus concentration. The solid black line at 0.5 relative ingot height indicates the transition from p-type conductivity to n-type conductivity.

3.4 Majority- and minority-carrier mobilities in compensated crystalline silicon

The mobility of electrons and holes in silicon is determined by the amount of scattering they experience. Sources of scattering include lattice scattering, impurity scattering and carrier-carrier scattering. Since the probability of a scattering process is determined by both the scattering cross-section and the density of scattering centers, an increase in either results in a decreased mobility. Compensated silicon in general and UMG-Si in particular contain much more dopants than standard silicon, and thus considerably higher amounts of impurities and carriers. A reduction of the carrier mobility in this kind of material is thus expected. However, no experimental data on mobilities in highly compensated silicon was available in the literature at the beginning of this work.

3.4.1 Block-cast multicrystalline silicon

The impact of compensation on the carrier mobilities in multicrystalline silicon is investigated on an mc-Si ingot made from upgraded metallurgical-grade (UMG) silicon. The exact boron and phosphorus concentrations in the melt are unknown. However, using the Scheil equation (Eq. 3.1) and the net doping concentrations obtained from electrochemical capacitance voltage

(ECV) measurements (see Sec. 2.2), the melt concentrations for both B and P are estimated to exceed 2×10^{17} cm^{-3}.

The majority-carrier mobilities (i.e. μ_p in p-type wafers and μ_n in n-type wafers) are determined through a combination of four-point probe resistivity ρ measurements and measurements of the equilibrium hole concentration p_0 (or, in n-type silicon, the equilibrium electron concentration n_0) using the ECV technique. The carrier mobilities then follow from $\mu_p = (p_0\, \rho\, q)^{-1}$ and $\mu_n = (n_0\, \rho\, q)^{-1}$, respectively, where q is the elementary charge.

As can be seen from Fig. 3.4, the hole mobility μ_p in the compensated p-type wafers is significantly lower than in non-compensated control samples, even though the resistivities are comparable. In addition, a further considerable decrease of the hole mobility in samples close to the transition region from p- to n-type silicon is observed. Beyond the transition point, however, the mobility increases again. Note that the majority-carriers are now electrons, which generally have higher mobilities than holes.

The pronounced reduction of μ_{maj} in the transition region can be attributed to a drastically reduced screening of the ionized scattering centers, caused by a significant decrease of the free carrier concentration. Based on the Brooks-Herring equation [39], we derive the following parameterization of the experimental data:

$$\mu_{\mathrm{maj}} = \frac{a_{\mathrm{maj}}}{(N_\mathrm{A} + N_\mathrm{D})\left(\ln \frac{b_{\mathrm{maj}}}{p_0} - 1\right)}, \tag{3.2}$$

where a_{maj} and b_{maj} are prefactors. The values for the boron concentration N_A, the phosphorus concentration N_D and net doping concentration p_0 are obtained from modeling the distribution of boron and phosphorus over the ingot height by using the hole concentrations p_0 obtained from the ECV measurements and the Scheil equation (Eq. 3.1). The result is plotted in Fig. 3.4 (dark red line), where $a_{\mathrm{maj}} = 4.3 \times 10^{20}$ (cm V s)$^{-1}$ and $b_{\mathrm{maj}} = 7.0 \times 10^{19}$ cm^{-3}. As can be seen, the agreement with the measured data of μ_{maj} is excellent.

The minority-carrier mobility (μ_n in p-type wafers and μ_p in n-type wafers) is determined through spatially resolved effective carrier lifetime τ_{eff} measurements on as-cut wafers using a microwave-detected photoconductance decay (MW-PCD) setup (see Sec. 2.1.2). Due to the saw damage, the surface recombination velocity S of such as-cut wafers can be expected to exceed 10^6 cm/s. Using a bias-light intensity of 30 mW/cm^2 and low laser power, the samples are kept in low-injection ($\Delta n \ll p_0$) and τ_{eff} is thus limited by the diffusion of minority-carriers to the surfaces (see Sec. 2.1.2).

These measurements are performed on wafers cut horizontally from the mc-Si ingot as well as on a wafer which was cut vertically from the ingot. As a result, the evolution of the minority-carrier mobility over almost the entire ingot height can be studied on a single wafer (the wafer was cut to 156×156 mm^2, while the ingot height was 195 mm). The resulting lifetime mapping is shown in Fig. 3.5(a). The red rectangular specifies the region over which an averaged linescan is drawn, shown in Fig. 3.5(b). In the linescan, the pronounced peak of τ_{eff} in the transition

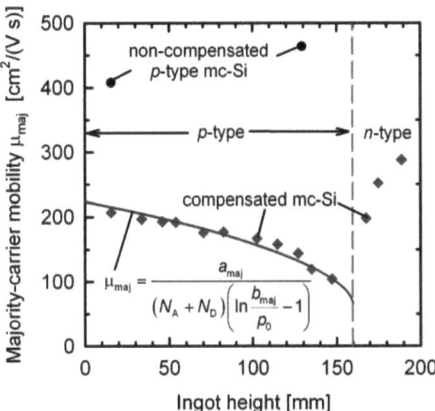

Figure 3.4: Majority-carrier mobilities μ_{maj} in a compensated mc-Si ingot, derived from four-point probe resistivity ρ measurements and determination of the carrier concentration p_0 (or n_0), plotted versus the ingot height (blue diamonds). The black circles give reference values obtained in non-compensated p-type mc-Si of similar resistivity. The transition from p- to n-type conductivity is indicated by the dashed green line.

region is clearly visible.

The lifetime image from Fig. 3.5(a) is subsequently converted into a minority-carrier mobility $\mu_{\min}$ mapping by using Eqs. 2.16 and 2.17, as shown in Fig. 3.6(a). It should be noted that an infinite bulk carrier lifetime τ_b is assumed for this calculation. Given that multicrystalline silicon is investigated, this assumption does not hold true, as is demonstrated by the far too high $\mu_{\min}$ values [> 1500 cm^2/(V s)] obtained for the edge regions. However, looking at Eq. 2.16, one finds that a finite bulk carrier lifetime will always yield a *smaller* diffusion coefficient $D_{\min}$, and accordingly a *smaller* carrier mobility. The values for the minority-carrier mobilities stated here thus pose an upper limit, further stressing all observed reductions.

An averaged linescan of the minority-carrier mobility $\mu_{\min}$ over the wafer (and thus the ingot height), indicated by the red rectangular in Fig. 3.6(a), is shown in Fig. 3.6(b) (blue diamonds). Due to the above-mentioned impact of a finite bulk carrier lifetime τ_b on the calculation of $D_{\min}$ and $\mu_{\min}$, the highly contaminated edge regions as well as the bottom (< 45 mm ingot height) and top (> 175 mm) of the ingot are excluded from the linescan, since these regions are known to contain increased amounts of metal impurities.

From 70 mm to 110 mm ingot height, the electron mobility is almost constant at $\mu_n = 580$ cm^2/(V s). This value is notably lower than that determined for non-compensated p-type mc-Si of similar resistivity [there, $\mu_n = 1073$ cm^2/(V s)]. At lower ingot height (between 45 mm and 70 mm), the minority-carrier mobility steeply increases, which is most likely due to

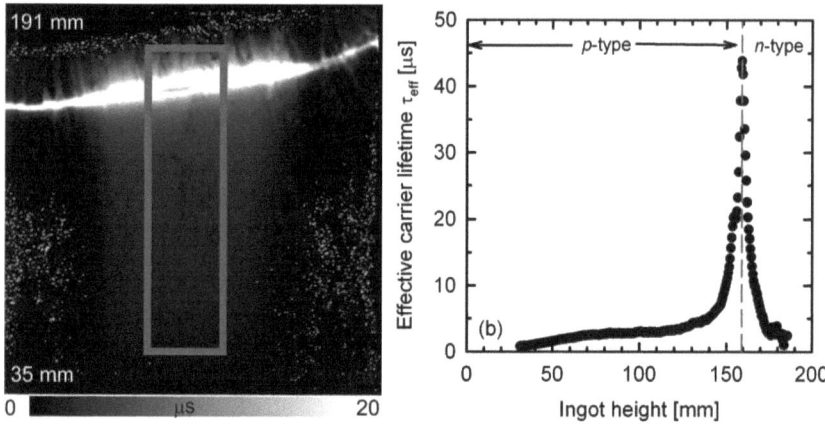

Figure 3.5: (a) Effective carrier lifetime τ_{eff} mapping of an as-cut wafer cut vertically from a compensated mc-Si ingot. The bottom edge of the wafer corresponds to an ingot height of 35 mm. The red rectangular marks the region over which an averaged linescan was drawn. (b) Averaged linescan of the effective lifetime τ_{eff} data shown in Fig. 3.5(a), plotted versus the ingot height. There is a sharp increase of τ_{eff} in the transition region from p- to n-type silicon (indicated by the dashed green line).

significantly reduced bulk lifetimes, as explained above. At ingot heights above 110 mm, μ_{min} starts to steadily decrease until reaching its minimum in the 'transition point' from p- to n-type silicon (indicated by the dashed green line). Conveniently, even though the n-type portion of the ingot is significantly smaller than the p-type region, the increase of the minority-carrier mobility (now μ_p) beyond the transition point is well visible.

With regard to the transition region, it must be noted that the assumption of low-level injection will at some point be violated (i.e. $\Delta n \approx p_0$ or even $\Delta n > p_0$). In that region, the minority-carrier diffusion coefficient D_{min} in Eq. 2.16 must be replaced by the ambipolar diffusion coefficient D_a, defined as $D_a = [(n+p)/(n/D_p + p/D_n)]$, with the electron concentration $n = n_0 + \Delta n$, hole concentration $p = p_0 + \Delta p$ as well as the diffusion coefficients for holes D_p and electrons D_n. Estimating that the excess carrier density Δn is below 5×10^{13} cm^{-3} under the bias light intensity of 30 mW/cm^2, the sample is in low-level injection when $p_0 > 5 \times 10^{14}$ cm^{-3}. Looking at the hole (and electron) concentrations obtained from the ECV measurements, the region where $p_0 < 5 \times 10^{14}$ cm^{-3} can be narrowed to be ± 10 mm (at most) around the transition point, as indicated by the yellow background in Fig. 3.6(b).

Analogously to Eq. 3.2, we parameterize the minority-carrier mobility μ_{min} by:

$$\mu_{\text{min}} = \frac{a_{\text{min}}}{(N_A + N_D)\left(\ln \frac{b_{\text{min}}}{p_0} - 1\right)}. \tag{3.3}$$

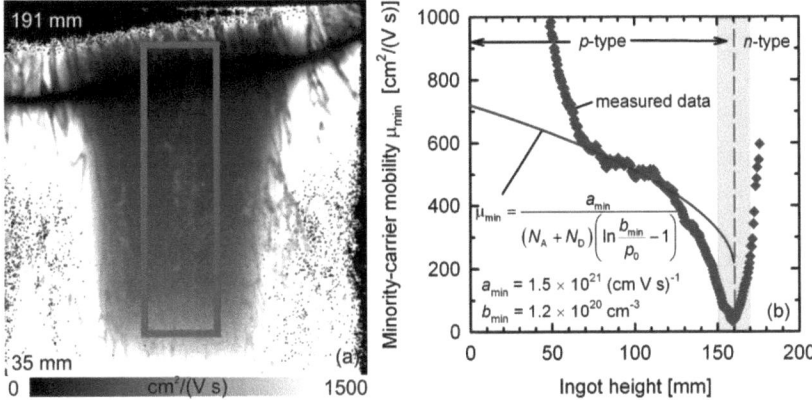

Figure 3.6: (a) Minority-carrier mobility $\mu_{\min}$ mapping of a wafer cut vertically from a compensated mc-Si ingot, calculated from the effective lifetime mapping shown in Fig. 3.5(a). The extremely high values at the edge of the wafer originate from falsely assuming an infinite bulk carrier lifetime τ_b. The red rectangular marks the region over which an averaged linescan was drawn. (b) Averaged linescan of the minority-carrier mobility $\mu_{\min}$ data shown in Fig. 3.6(a), plotted versus the ingot height. The dashed green line indicates the transition from p- to n-type silicon (and thus from electron to hole mobility).

Here, the prefactors are $b_{\min} = 1.2 \times 10^{20}$ cm^{-3} and $a_{\min} = 1.5 \times 10^{21}$ (cm V s)$^{-1}$. The result is plotted in Fig. 3.6(b) (dark red line). As can be seen, the parameterization yields very good agreement from 70 mm to 120 mm ingot height. The considerable deviation below 70 mm can, as mentioned before, most likely be attributed to a noticeably reduced bulk carrier lifetime. In addition, the decrease of the measured $\mu_{\min}$ values is much steeper than that predicted by the parameterization between 120 mm and 150 mm ingot height. This could be related to other so far unidentified scattering mechanisms, which become more important in that region and which are more effective for electrons (or minority-carriers in general) than for holes (or majority-carriers).

3.4.2 Monocrystalline Czochralski-grown silicon

To study the impact of compensation on the carrier mobility in monocrystalline silicon, samples from two Cz-Si ingots are investigated. As opposed to the UMG-Si used to make the mc-Si ingots from the last Section, high-purity feedstock was used for these two crystals.

In Ingot A, the boron and phosphorus concentrations added to the melt were $[B]_{\text{melt}} = [P]_{\text{melt}} = 3 \times 10^{16}$ cm^{-3}. Accordingly, the transition from p- to n-type conductivity is located at about 98% relative distance from the seed end. In Ingot B, the boron concentration added to the melt was $[B]_{melt} = 6 \times 10^{16}$ cm^{-3} while the phosphorus concentration added to the melt was $[P]_{melt}$ = 9×10^{16} cm^{-3}. The transition from p- to n-type conductivity is located at about 25% ingot height. As a result, Ingot A has mostly p-type conductivity while the bigger part of Ingot B has n-type conductivity.

As was the case for the mc-Si in the previous Section, the majority-carrier mobilities μ_{maj} are obtained from a combination of four-point-probe resistivity ρ measurements and the determination of the net doping concentration n_0 (or p_0) with the ECV technique according to $\mu_n = (n_0\ \rho\ q)^{-1}$ and $\mu_p = (p_0\ \rho\ q)^{-1}$, respectively (where q is the elementary charge).

When applied to non-compensated control samples, this method results in $\mu_p = (480 \pm 50)$ cm^2/(V s) in 1.3 Ω cm p-type Float zone silicon and $\mu_n = (1750 \pm 200)$ cm^2/(V s) in 1.5 Ω cm n-type Czochralski-grown silicon. These values are slightly higher than those derived from standard mobility models. For example, in the model of Klaassen [40, 41], which is a good parameterization of existing experimental data reported for non-compensated silicon, μ_p = 432 cm^2/(V s) in non-compensated 1.3 Ω cm p-type silicon, whereas μ_n = 1320 cm^2/(V s) in non-compensated 1.5 Ω cm n-type silicon. However, taking into account the scatter range of the experimental data in the literature, the agreement between our measurements and existing data is very good.

The minority-carrier mobility μ_{min} is obtained from effective carrier lifetime measurements on as-cut wafers using the MW-PCD technique. Measurements on non-compensated samples yield $\mu_n = (1180\pm40)$ cm^2/(V s) in 1.0 Ω cm p-type FZ-Si and $\mu_p = (424\pm20)$ cm^2/(V s) in 1.0 Ω cm n-type FZ-Si (compared to μ_n = 1060 cm^2/(V s) and μ_p = 440 cm^2/(V s) according to Klaassen's parameterization). Note that an infinite bulk lifetime τ_b is assumed in the analysis. Given that lifetimes down to 20 μs are measured in degraded low-resistivity samples from the seed end of the ingot, this assumption may result in an overestimation of the minority-carrier mobility by up to 20%.

The measured hole mobilities μ_p are depicted in Fig. 3.7, where Fig. 3.7(a) shows μ_p in p-type samples from Ingot A (majority-carrier mobility) and Fig. 3.7(b) shows μ_p in n-type samples from Ingot B (minority-carrier mobility), plotted versus the ingot height. The transition from p- to n-type conductivity is indicated by the dashed green line.

The hole mobilities in Ingot A are reduced by 25% to 60% when compared to hole mobilities in non-compensated Cz-Si. Note that the reduction is almost constant over the entire height of the ingot, with a slight additional decrease at the tail end. This trend, as well as the overall reduction, can be attributed to the increased amount of dopants in the material, which are ionized impurities. As a result, the scattering of free charge carriers is increased and the mobility is reduced. Accordingly, the largest reduction (40% to 60%) is observed at the tail end of the ingot, where the dopant concentrations are the highest.

In Ingot B, the hole mobility μ_p (shown in Fig. 3.7(b)) is even lower than in Ingot A, which can be explained by the higher dopant concentrations in the melt (and consequently in the

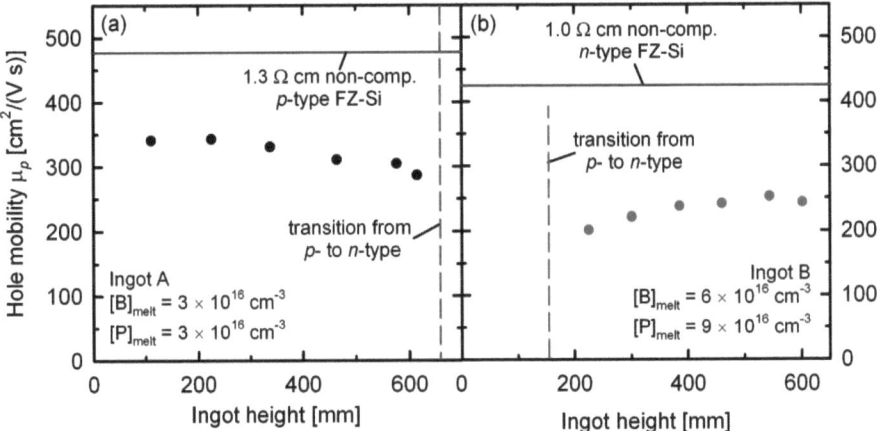

Figure 3.7: Hole mobility μ_p in two compensated Cz-Si ingots, plotted versus the ingot height. Apart from the overall reduction of μ_p in comparison with non-compensated Cz-Si, there is an additional decrease of the hole mobility in the vicinity of the transition region from p- to n-type conductivity (as indicated by the dashed green line.)

ingot). The material thus contains even more ionized impurities, which leads to even lower carrier mobilities. In addition, the hole mobility μ_p decreases further with decreasing distance from the transition region (as indicated by the dashed line). Note that this reduction cannot be attributed to a significant increase in the number of scattering centers, as was the case for Ingot A, since the transition region from p- to n-type conductivity in Ingot B already occurs at 25% ingot height, where the dopant concentrations are almost constant.

Instead the observed reduction can be explained by lack of screening of the ionized impurities, as was already explained in Sec. 3.4.1: while the sum of donor N_D and acceptor atoms N_A remains almost constant, the free carrier concentration, i.e. the difference between N_D and N_A, decreases significantly, resulting in a weakened screening of the ionized impurities and accordingly in increased scattering cross sections [39].

Unfortunately, a fit of the mobility data using the Brooks-Herring formula (Eqs. 3.2 and 3.3) is not possible, because no reliable data of the distribution of dopants in the ingot is available. While the net doping concentration and consequently the boron and the phosphorus concentrations could be very well described by the Scheil equation in Sec. 3.4.1, this is not the case in the studied Cz-Si. This might be due to evaporation of dopants from the melt during crystal growth.

Figure 3.8 depicts the measured electron mobilities μ_n in the two ingots. The data from Ingot A is plotted in Fig. 3.8(a), while the data from Ingot B is plotted in Fig. 3.8(b). As was already the case for the hole mobility, the electron mobility in Ingot A is significantly reduced

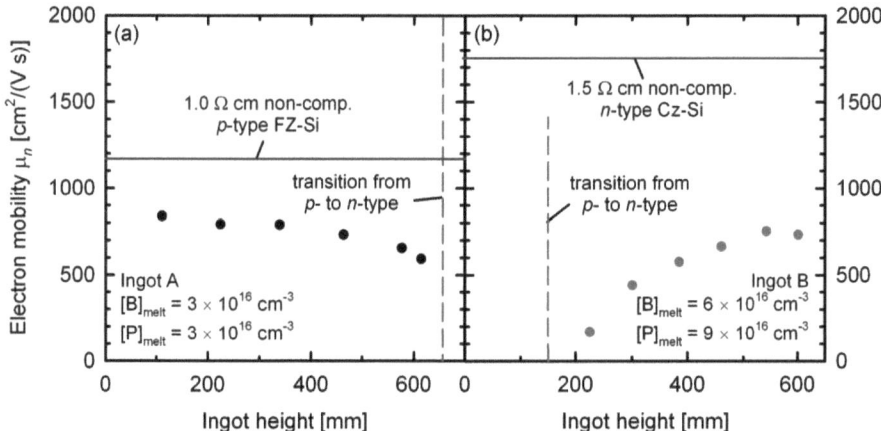

Figure 3.8: Electron mobility μ_n in two compensated Cz-Si ingots, plotted versus the ingot height. Apart from the overall reduction of μ_n in comparison with non-compensated Cz-Si, there is an additional decrease of the mobility in the vicinity of the transition region from p- to n-type conductivity (as indicated by the dashed green line.)

when compared to non-compensated silicon of similar resistivity. Again, the reduction is almost uniform over the bigger part of the ingot height, with an additional decrease at the tail end of the ingot. In Ingot B, the trend of the electron mobility is also very similar to that of the hole mobility: apart from the overall reduction of μ_n, there is an additional dramatic decrease close to the transition region.

Detailed data of the studied p-type Cz-Si wafers from Ingot A is summarized in Tab. 3.2. The error margins of μ_n stated in the table refer to variations over the wafer area. The total boron concentration N_A was determined through the iron-acceptor association time constant τ_{assoc} [42]. Table 3.3 compares the measured mobilities of the p-type samples with mobilities of non-compensated wafers of the same resistivity according to the parameterization of Klaassen [40, 41]. The values were obtained through a self-consistent calculation.

The experimental data obtained on the examined n-type Cz-Si wafers from Ingot B is summarized in Tab. 3.4. Also stated are the mobilities of non-compensated wafers of the same resistivity according to the model of Klaassen.

3.5 Chapter summary

In this Chapter, comprehensive data on majority- and minority-carrier mobilities in compensated mono- and multicrystalline silicon were presented for the first time. An overall reduction of the carrier mobilities of 25% to 50% was found in compensated Cz-Si, and of 45% to 55%

in compensated mc-Si. This reduction can mostly be attributed to the increased amounts of ionized impurities which act as scattering centers. In addition, a significant additional decrease of both μ_{maj} and μ_{min} was observed in the highly compensated transition region of all ingots. This decrease was explained by the decrease of the free carrier concentrations and the resulting weakened screening of the ionized dopant atoms. For the multicrystalline silicon ingots, a parameterization of the mobilities based on the Brooks-Herring formula resulted in excellent agreement with the experimental data of μ_{maj} and in good agreement for μ_{min}. This new mobility parameterization for compensated mc-Si will be useful for the numerical simulation of solar cells made on this novel type of silicon.

Table 3.1: Equilibrium segregation coefficients k_0 of the most common dopants and metal impurities in silicon. Data taken from [43].

Impurity	Equilibrium segregation coefficient k_0
B	8×10^{-1}
Ga	8×10^{-3}
Al	2×10^{-3}
In	4×10^{-4}
P	3.5×10^{-1}
O	1.25
C	7×10^{-2}
Fe	8×10^{-6}
Cu	4×10^{-4}
Ag	1×10^{-6}
Cr	1.1×10^{-5}

Table 3.2: Resistivities ρ of the studied p-type Cz-Si samples cut from Ingot A as measured by the 4-point-probe method, equilibrium hole concentrations $p_{0,\mathrm{ECV}}$ determined via ECV measurements, hole mobilities $\mu_{p,\mathrm{ECV}}$ calculated from ρ and $p_{0,\mathrm{ECV}}$, electron mobilities $\mu_{n,\mathrm{MWPCD}}$ determined from effective lifetime measurements on as-cut wafers, total boron concentrations N_A determined via the iron-acceptor association time constants τ_{assoc}, and total phosphorus concentrations N_D, calculated using N_A and $p_{0,\mathrm{ECV}}$.

Relative distance from seed end	ρ from 4-pp [Ω cm]	$p_{0,\mathrm{ECV}}$ from ECV [cm^{-3}]	$\mu_{p,\mathrm{ECV}}$ [cm^2/(V s)]	$\mu_{n,\mathrm{MWPCD}}$ [cm^2/(V s)]	N_A from τ_{assoc} [cm^{-3}]	$N_D = N_A - p_0$ [cm^{-3}]
0.16	1.45	1.3×10^{16}	326 ± 33	838 ± 32	1.5×10^{16}	0.2×10^{16}
0.32	1.50	1.2×10^{16}	329 ± 33	792 ± 33	1.8×10^{16}	0.6×10^{16}
0.48	1.70	1.1×10^{16}	320 ± 32	789 ± 34	1.9×10^{16}	0.8×10^{16}
0.66	2.1	9.0×10^{15}	323 ± 32	733 ± 34	1.9×10^{16}	1.0×10^{16}
0.82	3.7	6.1×10^{15}	272 ± 27	656 ± 34	2.1×10^{16}	1.5×10^{16}
0.88	5.1	4.5×10^{15}	271 ± 27	593 ± 36	2.3×10^{16}	1.8×10^{16}

Table 3.3: Resistivities ρ of the studied p-type Cz-Si samples cut from Ingot A as measured by the 4-point-probe method, measured hole mobilities $\mu_{p.\text{ECV}}$ from Tab. 3.2, hole mobilities $\mu_{p.\text{K}}$ in non-compensated silicon of resistivity ρ according to Klaassen's parameterization, measured electron mobilities $\mu_{n.\text{MWPCD}}$ from Tab. 3.2, and electron mobilities $\mu_{n.\text{K}}$ in non-compensated silicon of resistivity ρ according to Klaassen's parameterization.

Relative distance from seed end	ρ from 4-pp [Ω cm]	$\mu_{p.\text{ECV}}$ [cm^2/(V s)]	$\mu_{p.\text{K}}$ [cm^2/(V s)]	$\mu_{n.\text{MWPCD}}$ [cm^2/(V s)]	$\mu_{n.\text{K}}$ [cm^2/(V s)]
0.16	1.45	326 ± 33	435	838 ± 32	1118
0.32	1.50	329 ± 33	436	792 ± 33	1123
0.48	1.70	320 ± 32	440	789 ± 34	1145
0.66	2.1	323 ± 32	445	733 ± 34	1176
0.82	3.7	272 ± 27	456	656 ± 34	1248
0.88	5.1	271 ± 27	460	593 ± 36	1280

Table 3.4: Resistivities ρ of the studied n-type Cz-Si samples cut from Ingot B as measured by the 4-point-probe method, equilibrium electron concentrations $n_{0.\text{ECV}}$ determined via ECV measurements, electron mobilities $\mu_{n.\text{ECV}}$ calculated from ρ and $n_{0.\text{ECV}}$, electron mobilities $\mu_{n.\text{K}}$ in non-compensated silicon of resistivity ρ according to Klaassen's parameterization, hole mobilities $\mu_{p.\text{MWPCD}}$ determined from effective lifetime measurements on as-cut wafers, and hole mobilities $\mu_{p.\text{K}}$ in non-compensated silicon of resistivity ρ according to Klaassen's parameterization.

Relative distance from seed end	ρ from 4-pp [Ω cm]	$n_{0.\text{ECV}}$ from ECV [cm^{-3}]	$\mu_{n.\text{ECV}}$ [cm^2/(V s)]	$\mu_{n.\text{K}}$ [cm^2/(V s)]	$\mu_{p.\text{MWPCD}}$ [cm^2/(V s)]	$\mu_{p.\text{K}}$ [cm^2/(V s)]
0.37	4.9	5.6×10^{15}	285 ± 28	1385	201 ± 16	464
0.49	1.9	6.8×10^{15}	500 ± 50	1336	221 ± 13	453
0.62	1.05	1.0×10^{16}	595 ± 60	1290	237 ± 14	442
0.74	0.65	1.4×10^{16}	683 ± 69	1228	241 ± 11	428
0.88	0.43	2.0×10^{16}	747 ± 75	1185	252 ± 11	418
0.99	0.29	2.9×10^{16}	746 ± 75	1086	243 ± 17	396

4 Generation of boron-oxygen-related recombination centers

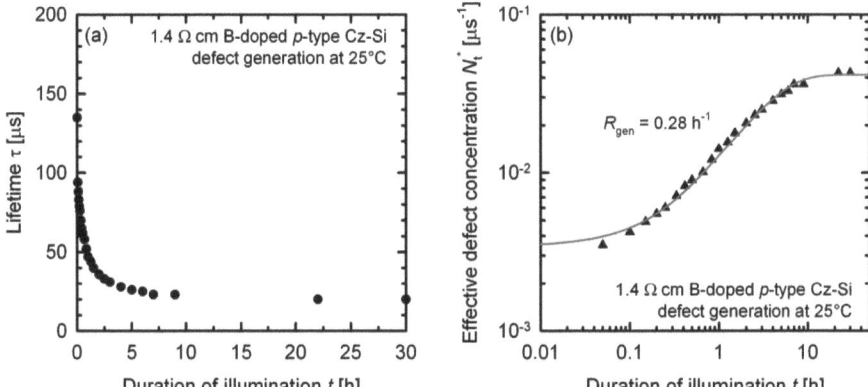

Figure 4.1: (a) Typical degradation of the lifetime τ in a 1.4 Ω cm boron-doped p-type Cz-Si wafer under illumination at room temperature due to the formation of boron-oxygen-related recombination centers. (b) Effective defect concentration N_t^*, derived from the lifetime data in Fig. 4.1(a), plotted on a double logarithmic scale. The data can be fitted by an exponential rise to maximum function (red line), which yields a defect generation rate constant of $R_{gen} = 0.28$ h^{-1}.

The carrier lifetime in boron-doped oxygen-rich crystalline silicon degrades under illumination at room temperature. This phenomenon has been known for a long time and has since then been intensely studied [2–19]. In those studies, light-induced degradation (LID) has been firmly linked to the simultaneous presence of both boron and oxygen [4]. Accordingly, the formation of boron-oxygen-related recombination centers has been proposed as the cause of LID.

The first Section of this Chapter briefly outlines the experimental results and theoretical approaches published before the beginning of this work. In the second Section, the formation kinetics as well as the extend of degradation are investigated for the first time in Czochralski-grown silicon (Cz-Si) doped with both boron and phosphorus. The defect concentration and the defect generation rate constant in compensated p-type Cz-Si are investigated as a function of the boron concentration N_A as well as of net doping concentration p_0. In addition, first experimental evidence of light-induced degradation in dopant-compensated n-type Cz-Si is presented.

An example of light-induced degradation in exclusively boron-doped Cz-Si is depicted in Fig. 4.1. In Fig. 4.1(a), the lifetime τ of a 1.4 Ω cm boron-doped p-type Cz-Si wafer, measured using the QSSPC technique and extracted at a fixed injection level of $\Delta n = 10^{15}$ cm^{-3}, is plotted versus the duration of illumination t at 25°C. Illumination is done at a light intensity of 10 mW/cm^2 using a halogen lamp. In the course of 30 hours, the lifetime decreases from initially 130 μs to its final value of 20 μs.

This decrease in lifetime corresponds to an increase of the concentration of recombination-active defects. Assuming that all other recombination channels, such as other defects or recombination at the surface, remain unchanged during light-induced degradation, the lifetime τ can be converted to an effective defect concentration N_t^* according to:

$$N_t^* = \frac{1}{\tau(t)} - \frac{1}{\tau_0}, \tag{4.1}$$

where $\tau(t)$ is the lifetime at time t and τ_0 is the initial lifetime before degradation.

Figure 4.1(b) depicts the effective defect concentration N_t^* derived from the lifetime data shown in Fig. 4.1(a). The increase of N_t^* as a function of time t can be described by an exponential function of the form $y = y_0 + a\,[1 - \exp(-R_{\text{gen}}t)]$, where R_{gen} is the defect generation rate constant. In the case shown, $R_{\text{gen}} = 0.28$ h^{-1} is obtained.

4.1 Review of previous experimental and theoretical work

Light-induced degradation (LID) of the energy conversion efficiency of solar cells made on boron-doped Czochralski-grown silicon (Cz-Si) was first observed by Fischer and Pschunder in 1973 [2]. After observing a loss of red spectral response, they concluded that the cause for degradation was a photon-induced decay of the carrier lifetime in the solar cell base. In addition, they found that a recovery of solar cell performance was possible through temperature treatments above 80°C. In particular, they reported a full restoration of the initial performance after annealing for 1 h at 200°C. The results obtained on solar cells were further supported by photoconductance decay measurements of the carrier lifetime. However, no explanation for the cause of these reversible changes in lifetime were offered.

Despite the far-reaching implications of this finding, no significant progress was made on the subject during the following 20 years. Eventually, in 1995, Knobloch et al. [3] presented high-efficiency solar cells fabricated on low-resistivity boron-doped Cz-Si, which suffered up to 10% relative loss in efficiency under illumination. Similar to Fischer and Pschunder [2], Knobloch et al. [3] also observed a full recovery of the solar cell performance after annealing; in their work at temperatures above 175°C. External quantum efficiency measurements again indicated that the degradation was caused by a degradation of the carrier lifetime in the bulk. Photoconductance decay measurements on unprocessed samples of the starting material confirmed this assumption as an exponential decay of the diffusion length was observed, which proceeded on the same time scale as the decay of the open-circuit voltage.

In 1997, Schmidt et al. presented the first systematic study of different Cz-Si materials doped with either boron, gallium or phosphorus [4]. Through this comparison, it was revealed that LID only occurs in B-doped Cz-Si, whereas the lifetime in Ga-doped Cz-Si and P-doped Cz-Si remains perfectly stable under illumination. In addition, they found that the extent of degradation increases with increasing boron concentration. In accordance with the previous results of Fischer and Pschunder [2] as well as of Knobloch et al. [3], Schmidt et al. also observed a full recovery of the lifetime after annealing. Furthermore, they demonstrated a full reversibility between the initial and the degraded state. Since the recovery behavior of the lifetime was found to correspond remarkably well with the annealing behavior of B_iO_i pairs, Schmidt et al. proposed that the recombination-active defect generated during LID was in fact the B_iO_i pair.

Further investigations by Glunz et al. [5] revealed that the effective concentration of the recombination-active defect, obtained from subtracting the inverse lifetime measured before and after light-induced degradation, i.e. $1/\tau_d - 1/\tau_0$, increased linearly with the boron concentration N_A. In addition, a strong dependence of the defect concentration on the interstitial oxygen concentration was found.

A detailed characterization of the electronic properties of the defect responsible for the lifetime degradation was performed in 1999 by Schmidt and Cuevas [6]. Through injection-dependent carrier lifetime measurements, which were analyzed using the Shockley-Read-Hall equation, they determined that the energy level of the defect was located between $E_V + 0.35$ eV and $E_C - 0.45$ eV, i.e. close to the middle of the silicon band-gap, and that the ratio of the electron and hole capture cross sections σ_n/σ_p was 10. This result was found to contradict the proposal that the B_iO_i pair was responsible for LID, since the energy level of the B_iO_i pair was known from deep level transient spectroscopy measurements to be at $E_t = E_C - 0.26$ eV. Instead, Schmidt and Cuevas suggested a defect of the type B_sO_{in}, i.e. a defect that is composed of one substitutional boron atom and n interstitial oxygen atoms.

After the linear dependence of the defect concentration on the boron concentration seemed assured, Glunz et al. examined the impact of the doping concentration on the defect generation rate constant $1/\tau_{gen}$ [7]. In that study, a quadratic dependence of $1/\tau_{gen}$ on the doping concentration was found. On the basis of this finding, Glunz et al. proposed that defect formation proceeded via a recombination-enhanced defect reaction [7]. However, no detailed suggestions concerning the nature of the defect or the actual transformation progress during LID were made.

In 2002, Bothe et al. reported on light-induced degradation in B-doped n-type silicon [11].

They investigated high-resistivity oxygen-rich Float zone silicon that was overcompensated by the formation of thermal donors. Light-induced degradation was observed both when the material had p- and n-type conductivity, however, the lifetime after degradation was much higher in the n-type state than in the p-type state.

In the same year, Hashigami et al. [12] presented time resolved measurements of current-induced degradation of the open-circuit voltage of solar cells, which revealed that the degradation actually proceeds in two distinct stages: a fast one that occurs on the time scale of seconds to minutes and a subsequent slow stage, which proceeds on a time scale of hours.

An unambiguous determination of the electronic properties was performed by Rein and Glunz via a combination of temperature- and injection-dependent lifetime spectroscopy (TDLS and IDLS) [13]. As a result, the energy level of the boron-oxygen-related recombination center was revealed to be at $E_\text{t} = E_\text{C} - 0.41$ eV, while the ratio of the electron and hole capture cross sections σ_n/σ_p was confirmed to be 9.3.

In 2004, Schmidt and Bothe proposed a new defect model for the boron-oxygen-related recombination center in crystalline silicon [14]. With regard to the experimental result that the defect concentration depends linearly on the boron concentration and quadratically on the interstitial oxygen concentration, they concluded that the defect is composed of one substitutional boron atom B_s and an interstitial oxygen dimer O_{2i}. Concerning the $B_\text{s}O_{2i}$ formation process, they proposed the capture of a fast-diffusing oxygen dimer by a substitutional boron atom. The diffusivity of the oxygen dimer was assumed to be increased by illumination due to a recombination-enhanced diffusion process.

In the same year, Adey et al. used density functional calculations to verify the proposed $B_\text{s}O_{2i}$-model [15]. They showed that two different configurations of the $B_\text{s}O_{2i}$ complex are stable in silicon, but only one of these configurations has an energy level in the silicon bandgap. Adey et al. [15] also demonstrated that an electrically stimulated enhanced diffusion of the oxygen dimer in silicon is possible.

Also in 2004, Bothe et al. performed carrier lifetime measurements to further investigate the two distinct stages of light-induced degradation [16]. Temperature-dependent measurements of the fast and slow defect generation rate constant showed no recognizable temperature dependence of the fast component of LID. The slow stage of defect formation, however, was found to be thermally activated with an activation energy of $E_\text{a} = 0.37$ eV [16].

In a more detailed study based on a larger number of samples in 2005, Bothe and Schmidt corrected their earlier finding and reported an activation energy of $E_\text{a} = 0.23$ eV for the formation of the fast forming recombination center [17]. In addition, similar to the slowly forming recombination center, the generation rate constant of the fast forming defect was found to increase quadratically with the doping concentration (and to be independent of the interstitial oxygen concentration). An analysis of the electronic properties yielded an energy level $E_\text{t} = E_\text{V} + 0.60$ eV and a capture cross-section ratio σ_n/σ_p of 100 [17].

In 2006, Herguth et al. demonstrated that illumination at elevated temperature (65°C to 160°C) resulted in a permanent recovery of the degraded open-circuit voltage of solar cells [20, 21]. That is, during illumination at elevated temperature the open-circuit voltage was observed to increase to its initial value before light-induced degradation and subsequently proved stable under illumination at 25°C for more than 100 hours. Additional experiments on lifetime samples confirmed that this permanent recovery was due to the recovery of the bulk carrier lifetime [20]. The process was found to be thermally activated with an activation energy of (0.64 ± 0.04) eV.

In 2007, Palmer, Bothe and Schmidt published a refined version of the B_sO_{2i}-model [19], in which the defect kinetics predicted a proportionality of R_{gen} on the product of the hole concentration p_0 and the total boron concentration N_A, i.e., $R_{gen} \propto p_0 \times N_A$. The dependence on p_0 followed from the requirement of the oxygen dimer to catch a hole to speed up its diffusivity, while the dependence on N_A was a consequence of the oxygen dimers bonding with a B_s atom. This prediction was found to be in excellent agreement with the existing experimental data measured on non-compensated p-type Cz-Si, since in exclusively B-doped Cz-Si $p_0 \times N_A = N_A^2$.

4.2 Light-induced degradation in dopant-compensated Cz-Si

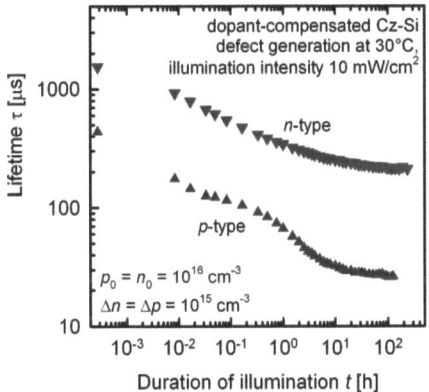

Figure 4.2: Comparison of the lifetime degradation in two dopant-compensated Cz-Si wafers, one p-type (blue triangles up) and one n-type (red triangles down), under illumination at 30°C and at a light intensity of 10 mW/cm², plotted on a double logarithmic scale. Both samples have a net doping concentration of 10^{16} cm^{-3} and the lifetime τ is extracted at a fixed excess carrier density of $\Delta n = \Delta p = 10^{15}$ cm^{-3}. While the time frame of the degradation is comparable for both samples, the shape of the degradation differs considerably.

Due to its dependence on boron, light-induced degradation is also expected in compensated silicon doped with both boron and phosphorus. The investigation of LID in compensated silicon allows for new insights into the defect composition and defect kinetics, since the net doping concentration p_0 and the boron concentration N_A can now be investigated separately. In addition, compensated silicon which contains more phosphorus than boron has n-type conductivity, which allows to investigate doping concentration and hole concentration separately.

Figure 4.2 depicts the carrier lifetime τ in two compensated samples doped with both boron and phosphorus. One of the samples has p-type conductivity (blue triangles up) while the other sample has n-type conductivity (red triangles down). The net doping concentration in both samples is $p_0 = n_0 = 10^{16}$ cm^{-3} and the lifetime is extracted at a fixed injection level of $\Delta n = \Delta p = 10^{15}$ cm^{-3}. Interestingly, the degradation proceeds in a similar time frame of 50 h to 100 h in both samples, despite the different conductivity type. On the other hand, the shapes of the degradation differ considerably.

4.2.1 Compensated *p*-type Cz-Si

In order to study LID in compensated p-type Cz-Si, a set of six wafers, all cut from the same ingot (Ingot A), is investigated. The resistivity ρ of the samples ranges from 1.45 Ω cm (close to the seed end) to 5.1 Ω cm (close to the tail end), whereas the substitutional boron concentration N_A varies between 1.5×10^{16} cm^{-3} and 2.3×10^{16} cm^{-3}. The interstitial oxygen concentration is in the range of $[O_i] = (10 \pm 2) \times 10^{17}$ cm^{-3} (at the seed end) to $[O_i] = (7 \pm 1) \times 10^{17}$ cm^{-3} (at the tail end). For further sample details see Tab. 3.2. Processing steps include acidic damage etching, RCA cleaning, phosphorus diffusion, removal of the n^+ layers from both sides of the wafer and surface passivation using plasma-enhanced chemical vapor deposited (PECVD) silicon nitride SiN$_x$ [44].

Figure 4.3(a) depicts the lifetime τ of three of these samples during illumination at 30°C and a light intensity of 10 mW/cm^2. The lifetime is extracted at a fixed injection level of $\Delta n = 0.1 \times p_0$. Similar to non-compensated p-type silicon, the lifetime both before LID and after complete defect formation decreases with decreasing resistivity ρ: the 5.1 Ω cm wafer (purple squares) has an initial lifetime of 1300 μs, whereas the 2.1 Ω cm wafer (yellow triangles down) has an initial lifetime of 560 μs and the 1.45 Ω cm wafer (green triangles up) has an initial lifetime of 365 μs. After complete defect formation, the lifetime of the 5.1 Ω cm sample is 59 μs, of the 2.1 Ω cm sample 32 μs and of the 1.45 Ω cm sample 20 μs.

Accordingly, the effective defect concentration N_t^*, depicted in Fig. 4.3(b), after complete defect formation increases with decreasing resistivity. In addition, Fig. 4.3(b) also shows that the defect generation rate constant R_{gen}, as determined from exponential fits of the data (solid lines), decreases with increasing resistivity. Note that the deviation of the exponential fits at the very early stage of LID is due to the fast forming recombination center [12, 16, 17].

In exclusively boron-doped Cz-Si, the effective defect concentration N_t^* was found to increase linearly with the boron concentration N_A and quadratically with the interstitial oxygen concentration $[O_i]$ [5, 14, 17, 18]. These findings led to the development of a defect model where the

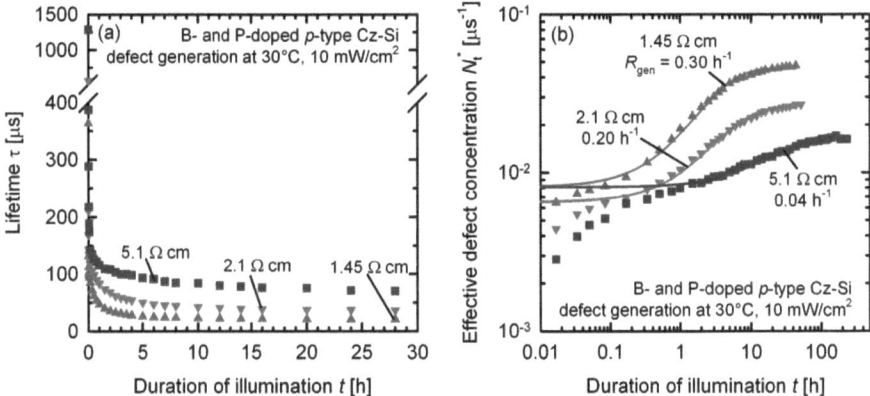

Figure 4.3: Time dependence of (a) the lifetime τ and (b) the effective defect concentration N_t^* during illumination at 30°C in three dopant-compensated p-type Cz-Si samples with different resistivities ρ. To determine the defect generation rate constant R_{gen}, an exponential function is fitted to the data of N_t^* (solid lines).

recombination center is composed of a substitutional boron atom B_s and an interstitial oxygen dimer O_{2i} to form a recombination-active B_sO_{2i} complex [14, 15, 19].

However, in exclusively B-doped Cz-Si, the boron concentration N_A equals the doping concentration p_0, and accordingly no distinction between N_A and p_0 is possible. In compensated p-type Cz-Si doped with both boron and phosphorus, on the other hand, the net doping concentration p_0 differs from the boron concentration N_A. Accordingly, such material allows to study the impact of p_0 and N_A separately.

Figure 4.4 shows the effective defect concentration N_t^* measured in the compensated p-type samples plotted versus both p_0 (red triangles up) and N_A (blue triangles down). Conveniently, the net doping concentration p_0 and the total boron concentration N_A in the present set of samples exhibit opposing trends: p_0 decreases with increasing distance from the seed end, while N_A increases with increasing distance from the seed end. As a result, an increase of N_t^* with increasing p_0 is observed while N_t^* decreases with increasing N_A.

Given that in all studies on exclusively B-doped Cz-Si an increase of N_t^* with increasing $N_A = p_0$ was observed, an opposing trend in compensated Cz-Si is difficult to explain. On the other hand, a linear increase of N_t^* with increasing p_0 is in excellent agreement with the existing experimental data on non-compensated Cz-Si.

At this point, it should be noted that (in non-compensated p-type Cz-Si) the defect concentration N_t^* has been found to depend quadratically on the interstitial oxygen concentration $[O_i]$ [5, 14, 17, 18]. In the present set of samples, $[O_i]$ decreases with increasing distance from the seed end of the ingot and accordingly with increasing N_A. As a result, it is possible that

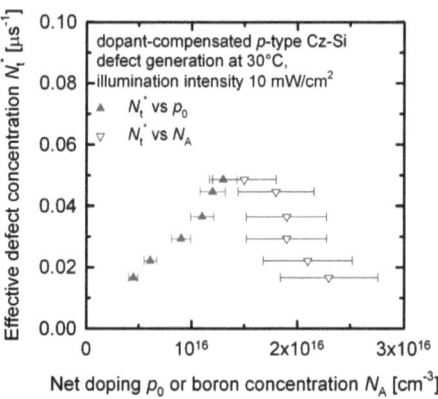

Figure 4.4: Effective defect concentration N_t^* in dopant-compensated p-type Cz-Si after complete degradation, plotted versus the net doping concentration p_0 (red triangles up) and the total boron concentration N_A (blue triangles down), respectively. N_t^* increases with increasing p_0 but decreases with increasing N_A. Given that in exclusively B-doped Cz-Si, N_t^* was found to increase linearly with $p_0 = N_A$, a decrease of N_t^* with increasing N_A in compensated Cz-Si is highly unlikely. Thus, these results strongly indicate that N_t^* actually depends on p_0 and not on N_A.

an increase of N_t^* due to the increase of N_A is counteracted by a simultaneous decrease of N_t^* due to the decrease of $[O_i]$, leading to a wrong conclusion regarding the dependence of N_t^* on N_A.

However, the interstitial oxygen concentration in Ingot A decreases from $[O_i] = (10 \pm 2) \times 10^{17}$ cm^{-3} to $[O_i] = (7 \pm 1) \times 10^{17}$ cm^{-3}. This decrease accounts for a decrease of N_t^* by a factor of 2. The actual observed decrease of N_t^* with increasing N_A (and decreasing $[O_i]$) is by a factor of 3, i.e. *higher* than the decrease caused by the decreasing interstitial oxygen concentration. However, since the boron concentration N_A simultaneously increases, the decrease in N_t^* should be *lower*. Accordingly, we conclude that the presented measurements indeed demonstrate that N_t^* actually depends on p_0. Note that this dependence is in contradiction to the B_sO_{2i}-model [14, 19].

With regard to the defect generation rate constant R_{gen}, the B_sO_{2i}-model predicts a proportionality of R_{gen} to the product of net doping and total boron concentration $p_0 \times N_A$ [19]. In non-compensated Cz-Si, R_{gen} was found to be proportional to N_A^2 [8, 9, 18], which of course equals $p_0 \times N_A$ in exclusively boron-doped silicon. Consequently, theory and experiment were found to agree in non-compensated p-type Cz-Si.

In order to verify this dependence in compensated p-type silicon, we compare the measured values of R_{gen} in the compensated p-type Cz-Si samples to R_{gen} values measured on non-compensated p-type Cz-Si (dashed line) in Fig. 4.5. When the defect generation rate constant

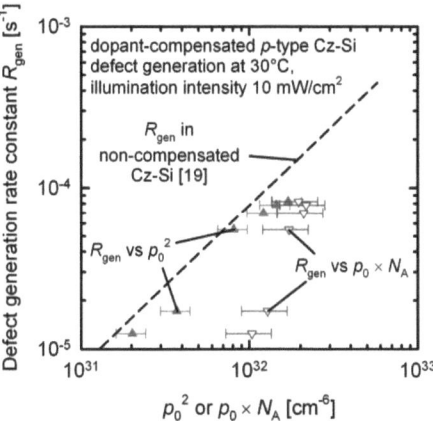

Figure 4.5: Defect generation rate constant R_{gen} determined in dopant-compensated p-type Cz-Si plotted on a double logarithmic scale versus the square of net doping concentration p_0^2 (red triangles up) and the product of the net doping and the total boron concentration $p_0 \times N_A$ (blue triangles down), respectively. The dashed line is a fit to existing data of R_{gen} in non-compensated Cz-Si, where $p_0 = N_A$ [19]. The agreement between compensated and non-compensated material is much better when R_{gen} is plotted versus p_0^2 than for the case where R_{gen} is plotted versus $p_0 \times N_A$.

is plotted as a function of $p_0 \times N_A$ (blue triangles down), there is a notable deviation between R_{gen} in the compensated samples and R_{gen} in the non-compensated wafers. However, when R_{gen} is plotted as a function of the square of the net doping concentration p_0^2 (red triangles up), the agreement with the data obtained on non-compensated silicon is good. Again, this new result obtained on compensated silicon cannot be explained with the state-of-the-art B_sO_{2i}-model [14, 19]

The new findings presented here thus call for a reassessment of the B_sO_{2i} defect model. An alternative defect model, which is capable of explaining our new results on compensated silicon, will be discussed in Chapter 7.

4.2.2 Compensated n-type Cz-Si

Compensated n-type Cz-Si provides an opportunity to further separate the parameters that might play a role in the formation of the boron-oxygen-related recombination center. In particular, the net doping concentration n_0 now differs from the hole concentration p. Studying light-induced degradation in compensated n-type Cz-Si is thus an important aspect in fully understanding the nature and formation kinetics of the defect.

The samples used in this work to study LID in compensated n-type Cz-Si were all cut from

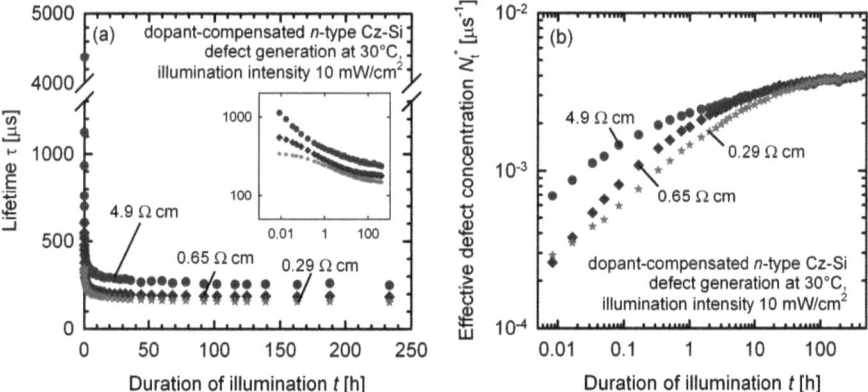

Figure 4.6: (a) Lifetime τ plotted versus duration of illumination t at 30°C and at a light intensity of 10 mW/cm² of three dopant-compensated n-type Cz-Si samples with different resistivities ρ. To give a better impression of the early stages of light-induced degradation, the inset shows the data plotted on a double logarithmic scale. (b) Effective defect concentration N_t^* calculated from the lifetime data shown in Fig. 4.6(a), plotted on a double logarithmic scale. The time dependence of N_t^* is quite complex and cannot be described by a simple exponential function.

the same ingot (Ingot B). The resistivity ρ of the samples ranges from 4.9 Ω cm to 0.29 Ω cm, whereas the interstitial oxygen concentration is in the range $[O_i] = (11 \pm 3) \times 10^{17}$ cm^{-3} (at the seed end) to $[O_i] = (7 \pm 1) \times 10^{17}$ cm^{-3} (at the tail end). No detailed information about the boron concentration N_A is available, however, the boron concentration in the melt was $[B]_{\text{melt}} = 6 \times 10^{16}$ cm^{-3}. For further sample details see Tab. 3.4. The samples underwent a P-diffusion step and are passivated by PECVD-SiN$_x$. Photoconductance decay lifetime measurements are performed using a Sinton Instruments WCT-120 setup.

Figure 4.6(a) shows the evolution of the lifetime τ (extracted at a fixed injection level of $\Delta p = 0.1 \times n_0$) over time t in three compensated n-type samples with resistivities of 4.9 Ω cm (red circles), 0.65 Ω cm (blue diamonds) and 0.29 Ω cm (cyan stars), respectively. For a better impression of the early stages of degradation, the same data is plotted as an inset on a double logarithmic scale. As can be seen, there is a noticeable difference in the shape of lifetime reduction during the first few minutes of LID in the 4.9 Ω cm and the 0.29 Ω cm sample.

The time dependence of the effective defect concentration N_t^*, depicted in Fig. 4.6(b), is found to be quite complex and cannot be described by a simple exponential function. Interestingly, such a complex behavior may be explained by a dependence of the defect generation rate constant R_{gen} on the hole concentration. Since holes are minority carriers in n-type silicon, their concentration depends on the minority-carrier lifetime, which in turn depends on the concentration of recombination-active defects. During the course of light-induced degradation, the defect generation rate constant R_{gen} would thus decrease with increasing defect concentration N_t^*, resulting in a complex expression for the time dependence of N_t^*.

In addition, the final lifetime τ_d after complete degradation decreases with decreasing resistivity, and accordingly with increasing net doping concentration n_0. When calculating the effective defect concentration N_t^*, however, one finds very similar N_t^* values for all samples [see Fig. 4.6(b)]. Consequently, no clear dependence of N_t^* on n_0 is observed. Looking at the interstitial oxygen concentration $[O_i]$ in the samples, a decrease of $[O_i]$ with increasing n_0 is found. It is thus also conceivable that the increase of N_t^* due to the increase of n_0 might be counterbalanced by the decrease of $[O_i]$. More detailed studies are needed to clarify this point.

4.3 Chapter summary

This Chapter presented for the first time an in-depth experimental study on light-induced degradation in Czochralski-grown silicon doped with both boron and phosphorus. The defect generation was found to proceed in comparable time intervals in both compensated p- and n-type Cz-Si. However, the shape of the degradation in compensated n-type Cz-Si differs considerably from that in p-type Cz-Si.

The effective defect concentration N_t^* was investigated in compensated p-type Cz-Si samples of varying net doping concentrations p_0 and total boron concentrations N_A, leading to the pivotal result that N_t^* increases proportionally with increasing p_0 (rather than with N_A). In addition, the defect generation rate constant R_{gen} was studied for the first time in compensated silicon. It was found that it increases proportionally with the square of the hole concentration p_0^2.

Both experimental findings are not consistent with the state-of-the-art B_sO_{2i} defect model, which hence needs to be reassessed.

5 Annihilation of boron-oxygen-related recombination centers

The boron-oxygen-related recombination center is annihilated by a short anneal in the dark, leading to a recovery of the carrier lifetime [4,5]. However, this state is not stable and renewed illumination results in renewed generation of the recombination-active defect. This Chapter presents carrier lifetime studies on a variety of compensated p- and n-type samples that are doped with both boron and phosphorus, which improve the understanding of the annihilation kinetics. The defect annihilation rate constant R_{ann} is investigated as a function of the net doping concentration p_0 and n_0, respectively, and additionally as a function of total boron concentration N_A in the p-type samples.

5.1 Exclusively boron-doped p-type Cz-Si

An example of the defect annihilation in exclusively boron-doped Cz-Si is shown in Fig. 5.1. Three B-doped p-type samples with different resistivities ρ are annealed at 140°C in the dark on a hot plate. Figure 5.1(a) shows the measured lifetime τ versus time t at 140°C. Note that for each lifetime measurement the samples are removed from the hot plate. The lifetime is then measured at 29°C using the QSSPC technique and extracted at a fixed injection level of $\Delta n = 0.1 \times p_0$. Figure 5.1(b) shows the effective defect concentration N_t^* derived from the lifetime data shown in Fig. 5.1(a).

Both the lifetime in the degraded state τ_d and the lifetime after defect annihilation τ_0 increase with increasing resistivity. This is due to the dependence of the defect concentration on the doping concentration, as explained in Section 4.2.1. In addition, the annihilation proceeds at different rates in the three different samples. The lifetime in the 4.3 Ω cm sample (blue diamonds) saturates after approximately 3000 s, whereas the full recovery of the lifetime in the 0.40 Ω cm sample (black circles) takes more than 20000 s.

In order to determine the annihilation rate constant R_{ann}, an exponential function of the form $y = y_0 + a\exp\left(-R_{\mathrm{ann}}t\right)$ is fitted to the data in Fig. 5.1(b) (solid lines). As a result, we determine annihilation rate constants $R_{\mathrm{ann}} = 4.59 \times 10^{-4}$ s^{-1} for the 0.40 Ω cm sample, $R_{\mathrm{ann}} = 1.51 \times 10^{-3}$ s^{-1} for the 1.2 Ω cm sample and $R_{\mathrm{ann}} = 4.55 \times 10^{-3}$ s^{-1} for the 4.3 Ω cm sample. The annihilation rate constant of the 4.3 Ω cm sample is thus one order of magnitude higher

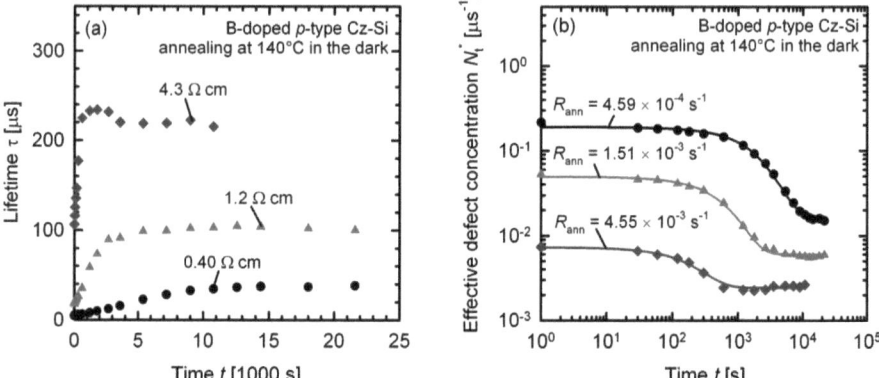

Figure 5.1: (a) Measured lifetime τ plotted versus time t at 140°C in darkness of three B-doped p-type Cz-Si samples of different resistivities. (b) Effective defect concentration N_t^* derived from the lifetime data shown in Fig. 5.1(a), plotted versus time t at 140°C in the dark on a double logarithmic scale. In order to determine the defect annihilation rate constant R_ann, an exponential function of the form $y = y_0 + a\exp\left(-R_\mathrm{ann}t\right)$ is fitted to the data.

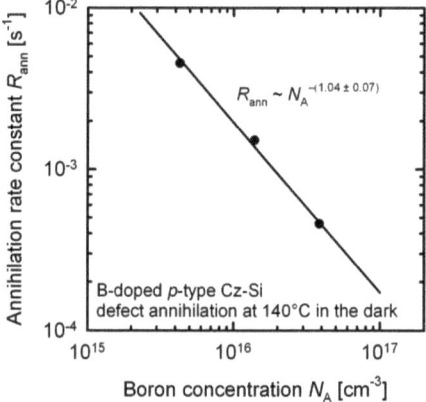

Figure 5.2: Annihilation rate constants R_ann in exclusively B-doped p-type Cz-Si determined at 140°C in darkness plotted versus the boron concentration N_A on a double logarithmic scale. The decrease of R_ann with increasing N_A can be described by a power law of the form $R_\mathrm{ann} \propto N_\mathrm{A}^{-(1.04\pm0.07)}$.

than that of the 0.40 Ω cm sample.

In Fig. 5.2, the annihilation rate constants R_ann determined in Fig. 5.1 are plotted versus the boron concentration N_A on a double logarithmic scale. The linear dependence on the double logarithmic scale, indicated by the solid line, corresponds to a power law with an exponent of $-(1.04 \pm 0.07)$, i.e. $R_{\mathrm{ann}} \propto N_A^{-(1.04 \pm 0.07)}$. Note that even though the studied set of samples is very limited, this result is in accordance with previously published data [10].

5.2 Boron- and phosphorus-doped Cz-Si

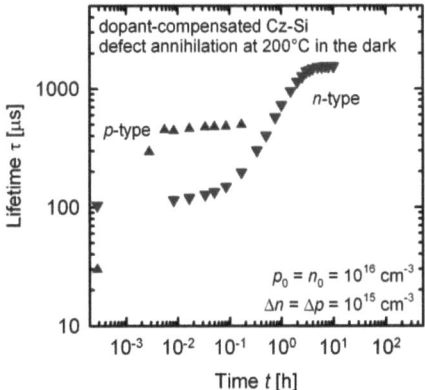

Figure 5.3: Lifetime τ plotted versus time t at 200°C in darkness of two dopant-compensated Cz-Si samples (doped with boron and phosphorus). Both samples have a net doping concentration of $p_0 = n_0 = 10^{16}$ cm^{-3} and the lifetime was extracted at a fixed excess carrier density of $\Delta n = \Delta p = 10^{15}$ cm^{-3}. Defect annihilation in the n-type sample (red triangles down) takes 1000 times longer than in the p-type sample (blue triangles up).

Figure 5.3 shows a comparison of the defect annihilation at 200°C in darkness for two compensated Cz-Si samples, one p- and one n-type. Both samples are doped with boron and phosphorus and have a similar net doping concentration of $p_0 = n_0 = 10^{16}$ cm^{-3}. The lifetime τ is extracted at a fixed injection level of $\Delta n = \Delta p = 10^{15}$ cm^{-3}. The defect annihilation in the p-type sample (blue triangles up) is completed after 30 to 60 s. Surprisingly, in the n-type sample (red triangles down) complete defect annihilation takes more than 10 hours.

5.2.1 Boron- and phosphorus-doped p-type Cz-Si

Two different sets of boron- and phosphorus-doped p-type Cz-Si samples are investigated. The first set consists of two wafers that were cut from two different ingots (Ingot 44 and Ingot 45). The sample from Ingot 44 has a resistivity ρ of 0.52 Ω cm, while the sample from Ingot 45 has a resistivity of 1.20 Ω cm. The net doping concentration p_0 is determined using the electrochemical capacitance voltage (ECV) technique, which yields $p_0 = 1.5 \times 10^{16}$ cm^{-3} for

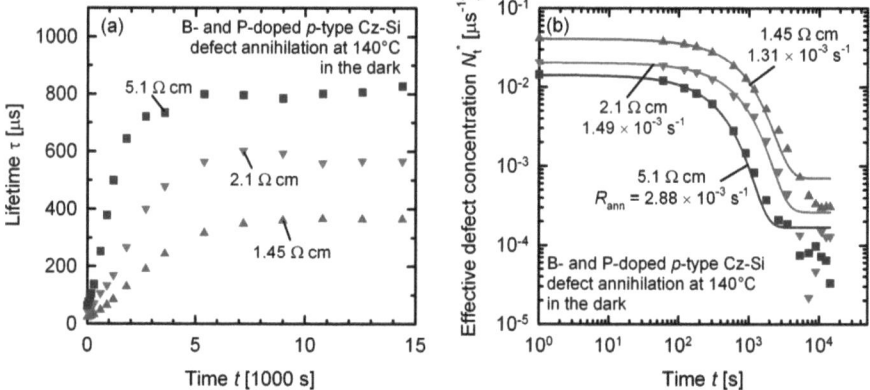

Figure 5.4: (a) Measured lifetime τ plotted versus time t at 140°C in darkness of three compensated p-type Cz-Si samples of different resistivities. The lifetime after completed defect annihilation increases with increasing resistivity. (b) Effective defect concentration N_t^* derived from the lifetime data shown in Fig. 5.4(a), plotted versus time t at 140°C in the dark on a double logarithmic scale. In order to determine the defect annihilation rate constant R_{ann}, an exponential function is fitted to the data (solid lines).

the 1.20 Ω cm sample (#45) and $p_0 = 4.1 \times 10^{16}$ cm^{-3} for the 0.52 Ω cm sample (#44). The boron concentration N_A is measured via the iron-acceptor association time constant τ_{assoc} [42] to be $N_A = 3.0 \times 10^{16}$ cm^{-3} for sample #45 and $N_A = 9.2 \times 10^{16}$ cm^{-3} for sample #44. The interstitial oxygen concentration [O$_i$] in both wafers is between 9.4×10^{17} cm^{-3} and 1.1×10^{18} cm^{-3}. The second set of samples consists of six wafers that were all cut from the same ingot (Ingot A, for sample details see Tab. 3.2).

An example of the defect annihilation in three of the compensated p-type Cz-Si samples from Ingot A is shown in Fig. 5.4. Figure 5.4(a) shows the lifetime τ, measured at a fixed injection level of $\Delta n = 0.1 \times p_0$, plotted versus the time t at 140°C in darkness, while Fig. 5.4(b) shows the effective defect concentration N_t^* derived from the lifetime data shown in Fig. 5.4(a). The resistivities of the samples are 5.1 Ω cm (purple squares), 2.1 Ω cm (yellow triangles down) and 1.45 Ω cm (green triangles up), respectively.

Similar to non-compensated p-type Cz-Si, the final lifetime after defect annihilation τ_0 noticeably increases with increasing resistivity. However, this increase in lifetime does not correspond to a similar decrease of the effective defect concentration N_t^*. Additionally, the data of N_t^* scatters strongly at the end of the annihilation process due to the fact that the values of N_t^* are comparatively small during that phase. In Fig. 5.4(b), an exponential function of the form $y = y_0 + a \exp\left(-R_{\mathrm{ann}}t\right)$ is fitted to the data in order to determine the defect annihilation rate constant R_{ann}. As can be seen, R_{ann} increases with increasing resistivity.

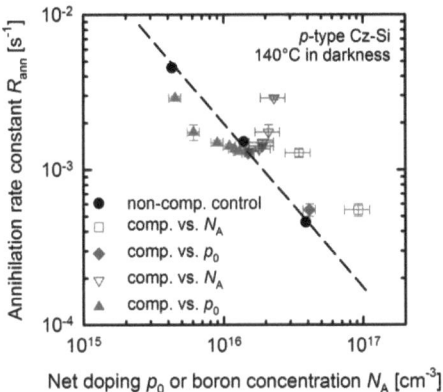

Figure 5.5: Annihilation rate constants R_{ann} in dopant-compensated p-type Cz-Si determined at 140°C in darkness plotted versus the net doping concentration p_0 (filled red triangles up and filled green diamonds) and the total boron concentration N_A (open pink squares and open blue triangles down), respectively, on a double logarithmic scale. When compared to data obtained on non-compensated p-type Cz-Si (black circles, dashed line), the agreement is much better when R_{ann} is plotted versus p_0.

Figure 5.5 shows the annihilation rate constants R_{ann} measured at 140°C in the dark of all investigated samples as a function of net doping concentration p_0 (filled red triangles up and filled green diamonds) and substitutional boron concentration N_A (open blue triangles down and open pink squares), respectively. As a reference, the R_{ann} values measured on non-compensated p-type Cz-Si shown in Fig. 5.2 are also included (black circles). When R_{ann} is plotted versus N_A (open symbols), there is considerable deviation from the data obtained on non-compensated Cz-Si. However, when the annihilation rate constant is plotted versus p_0 (filled symbols), the agreement between compensated and non-compensated samples is much better.

This is especially obvious for the six samples from Ingot A (triangles in Fig. 5.5): when R_{ann} is plotted versus the net doping concentration p_0, R_{ann} decreases with increasing p_0, as was the case for the non-compensated wafers. However, when R_{ann} is plotted versus the boron concentration N_A, there is an increase of R_{ann} with increasing N_A, which is contrary to the results obtained on non-compensated Cz-Si.

The experimental result that the defect annihilation rate constant R_{ann} is inversely proportional to the hole concentration p_0 is an important new finding with regard to identifying the kinetics of defect annihilation. In particular, the inverse dependence of R_{ann} on p_0 cannot be explained with the state-of-the-art defect model, as will be discussed in Chapter 7.

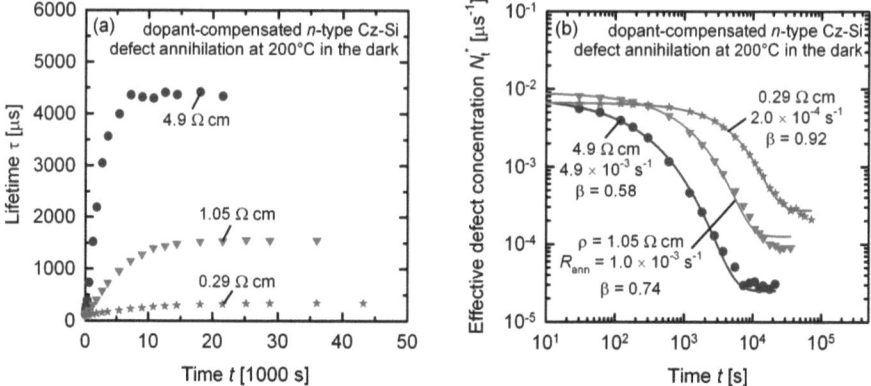

Figure 5.6: (a) Lifetime τ plotted versus time t at 200°C in darkness of three dopant-compensated n-type Cz-Si samples with different resistivities. The lifetime after complete defect annihilation increases with increasing resistivity. (b) Effective defect concentration N_t^* derived from the lifetime data shown in Fig. 5.6(a), plotted versus time t at 200°C in the dark on a double logarithmic scale. In order to determine the defect annihilation rate constant R_{ann}, a stretched exponential function of the form $y = \exp\left(-R_{\mathrm{ann}}t\right)^{\beta}$ was fitted to the data (solid lines).

5.2.2 Boron- and phosphorus-doped n-type Cz-Si

To study the defect annihilation in compensated n-type Cz-Si, a set of six samples, all cut from the same ingot (Ingot B), is used. The resistivities ρ are between 0.29 Ω cm and 4.9 Ω cm. For more sample details see Tab. 3.4. The samples are passivated using PECVD-SiN$_x$ and underwent a phosphorus diffusion (the resulting n^+ layers on both sides of the wafer were removed before surface passivation). The lifetime is measured via PCD measurements using a Sinton Instruments WCT-120 setup and are extracted at a fixed injection level of $\Delta p = 0.1 \times n_0$.

The time dependence of the lifetime τ in compensated n-type samples during defect annihilation at 200°C in the dark is shown in Fig. 5.6(a). The resistivities are 4.9 Ω cm (red circles), 1.05 Ω cm (yellow triangles down) and 0.29 Ω cm (cyan stars), respectively. Note that the annihilation takes up to 43000 s (i.e., 12 h), whereas in p-type silicon defect annihilation at 200°C is complete after 30 to 60 s (see Fig. 5.3). In addition, we find that the lifetime in compensated n-type Cz-Si after defect annihilation is considerably lower than in exclusively P-doped n-type Cz-Si of similar net doping concentration. For example, the 4.9 Ω cm sample recovers to $\tau = 4.5$ ms, whereas 12 ms are measured on a non-compensated 4.5 Ω cm control sample (with similar passivation).

Figure 5.6(b) shows the effective defect concentration N_t^* derived from the lifetime data shown in Fig. 5.6(a), plotted versus time t at 200°C in the dark on a double logarithmic scale. To determine the annihilation rate constant R_{ann}, a stretched exponential function of the form

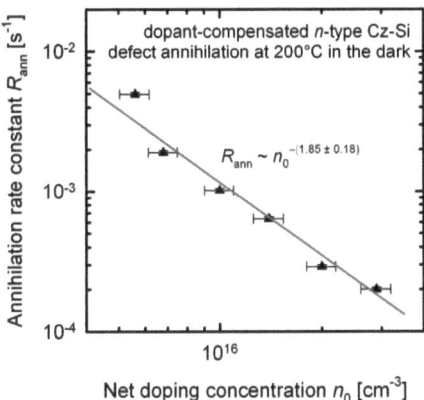

Figure 5.7: Annihilation rate constants R_{ann} measured in dopant-compensated n-type Cz-Si samples at 200°C in darkness plotted versus the net doping concentration n_0 on a double logarithmic scale. The decrease of R_{ann} with increasing n_0 can be described by a power law of the form $R_{\mathrm{ann}} \propto n_0^{-(1.85\pm 0.18)}$.

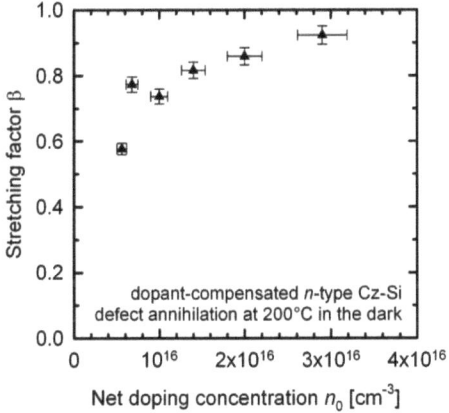

Figure 5.8: Stretching factor β obtained from the exponential fits in Fig. 5.6(b) plotted versus the net doping concentration n_0. β increases with increasing n_0, approaching 1. A stretching factor β of 1 corresponds to a single exponential decay function, which describes the defect annihilation behavior in B-doped p-type Cz-Si.

$y = \exp\left(-R_{\mathrm{ann}} t\right)^\beta$ is fitted to the experimental data (solid lines).

The obtained values of R_{ann} are plotted versus the net doping concentration n_0 in Fig. 5.7 on a double logarithmic scale. The observed decrease of R_{ann} with increasing n_0 can be fitted by a power law $R_\mathrm{ann} \propto n_0^{-(1.85\pm0.18)}$. Figure 5.8 shows the stretching factor β [originating from the fits in Fig. 5.6(b)] plotted versus the net doping concentration n_0. The stretching factor decreases with decreasing net doping concentration from $\beta = 0.92$ at $n_0 = 2.9 \times 10^{16}$ cm^{-3} to $\beta = 0.58$ at $n_0 = 5.6 \times 10^{15}$ cm^{-3}. Looking at other characteristics of compensated silicon, β also decreases with decreasing total boron concentration N_A and decreasing total phosphorus concentration N_D. On the other hand, β decreases with increasing compensation ratio $R_C = |(N_\mathrm{A} + N_\mathrm{D})/(N_\mathrm{A} - N_\mathrm{D})|$.

5.3 Chapter summary

In this Chapter, we have presented experimental results of the annihilation kinetics of the boron-oxygen-related recombination center in compensated Cz-Si doped with both boron and phosphorus. A surprising new finding was that the reaction speed of the defect annihilation is reduced by up to three orders of magnitude in compensated n-type Cz-Si when compared to compensated p-type Cz-Si of similar net doping concentration p_0 (and comparable total boron concentration N_A). Investigation of compensated p-type Cz-Si with varying net doping concentration p_0 and varying total boron concentration N_A revealed the pivotal finding that the annihilation rate constant R_ann in compensated p-type Cz-Si is inversely proportional to the net doping concentration p_0 (rather than the total boron concentration N_A). This has important consequences concerning the defect model (see Chapter 7).

In addition, the defect annihilation in compensated n-type Cz-Si was observed to exhibit stretched exponential characteristics, where the stretching factor β increased with increasing net doping concentration n_0. Additionally, an inverse dependence of the defect annihilation rate constant R_ann on n_0 was found, which was described by a power law $R_\mathrm{ann} \propto n_0^{-(1.85\pm0.18)}$.

Combined with the results presented in Chapter 4, these new experimental findings build the foundation for developing a novel defect model explaining both the degradation and annihilation in boron- and oxygen-containing silicon.

6 Permanent deactivation of boron-oxygen-related recombination centers

As has been shown in Chapter 4, the carrier lifetime in boron-doped Czochralski-grown silicon (Cz-Si) degrades under illumination at room temperature due to the formation of a recombination-active boron-oxygen-related defect. In Chapter 5, it was shown that this defect can be annihilated by annealing in the dark. However, this annihilated state is unstable under renewed illumination.

By contrast, Herguth et al. [20, 21] demonstrated that illumination at elevated temperature resulted in a recovery of the carrier lifetime and that this recovered lifetime was stable under subsequent illumination at room temperature. An example of this lifetime recovery is shown in Fig. 6.1, where the difference of the lifetime evolution in a 1.0 Ω cm B-doped p-type Cz-Si wafer during illumination at 25°C and 10 mW/cm^2 light intensity as well as during illumination at 185°C and 100 mW/cm^2 light intensity is depicted. Illumination is done using a halogen lamp, the intensity of which is adjusted with a calibrated solar cell. During illumination at elevated temperature, the sample is annealed on a hot plate that is set at 185°C. Lifetimes are measured using the QSSPC technique and extracted at a fixed injection level of $\Delta n = 10^{15}$ cm^{-3}. For each lifetime measurement the sample is removed from the hot plate. The temperature during the lifetime measurement is 29°C.

At $t = 0$, all boron-oxygen-related defects are annihilated by a 10 min anneal at 200°C in the dark, as described in Chapter 5, and the lifetime is accordingly high ($\tau_0 = 165$ μs). Under illumination at room temperature, light-induced degradation due to the generation of boron-oxygen-related recombination centers is observed (see Chapter 4). After only 1 minute of illumination, the lifetime drops to 78 μs, which is the so-called 'fast stage' of degradation [12,17]. Subsequently, τ decreases at a slower rate, until reaching its saturation value τ_d. Note that Fig. 6.1 does not actually depict τ_d, as complete degradation takes approximately 10 hours in 1.0 Ω cm Cz-Si.

As indicated by the arrow in Fig. 6.1, the same sample is again annealed at 200°C in darkness for 10 minutes before it is then placed under illumination at 185°C and at a light intensity of 100 mW/cm^2. During the first minutes, very fast light-induced degradation, which

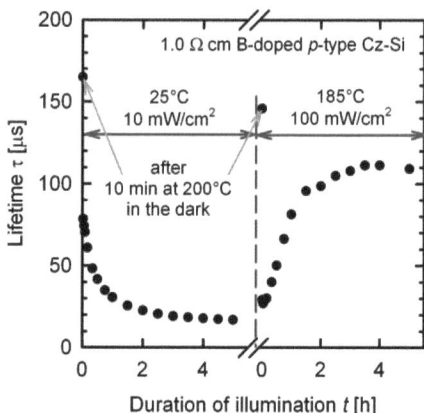

Figure 6.1: Typical evolution of the carrier lifetime τ in a 1.0 Ω cm B-doped p-type Cz-Si sample under illumination (after defect annihilation at 200°C in darkness). At room temperature (25°C), the carrier lifetime decreases (left of dashed blue line). At 185°C, τ initially decreases very rapidly but subsequently increases until it saturates (right of dashed blue line). After illumination at 185°C, τ is stable under subsequent illumination at room temperature.

is accelerated by the elevated temperature, is observed. Subsequently, however, the carrier lifetime recovers as the boron-oxygen complexes are deactivated. After 5 hours of illumination at 185°C, the wafer is removed from the hot plate. After this treatment, illumination at room temperature has no further effect on the carrier lifetime, suggesting a permanent deactivation of the boron-oxygen-related recombination center.

At the beginning of this work, very little was known about the permanent deactivation of the boron-oxygen-related defect. Herguth *et al.* reported that the process was thermally activated with an activation energy of $E_a = 0.7$ eV and that the deactivation could be observed both in lifetime samples and in fully processed solar cells [20, 21]. However, no detailed studies apart from the temperature dependence had been made.

This work presents the first comprehensive analysis of the permanent deactivation and it is shown that a large number of parameters affect the deactivation process. Particular processing steps such as a phosphorus diffusion or deposition of a PECVD silicon nitride layer can accelerate the deactivation. In addition, material properties such as the doping concentration, the interstitial oxygen concentration or the presence of thermal donors are found to have an impact on the permanent deactivation process. The results of these studies are presented in the first part of this Chapter.

The second part of this Chapter investigates the stability of the deactivated state. It is found that prolonged illumination at elevated temperature or extended annealing in darkness result in a destabilization of the deactivated state, leading to renewed degradation of the lifetime under illumination at room temperature. Finally, the deactivation procedure is applied to compensated n-type Cz-Si, which contains both boron and phosphorus. While a recovery of the lifetime under illumination at elevated temperature is observed, this lifetime is found to be unstable under illumination at room temperature, suggesting that a permanent deactivation of the BO defect is not possible in compensated n-type Cz-Si.

6.1 Deactivation of the BO complex in *p*-type Cz-Si

6.1.1 Impact of phosphorus diffusion

Since the majority of presently fabricated solar cells are made on boron-doped p-type silicon, a phosphorus diffusion is the most common process for emitter formation. Conveniently, a P-diffusion also removes fast-diffusing metal impurities from the bulk, which often results in improved carrier lifetimes. As a result, P-diffusions may also be applied to lifetime samples. However, for most lifetime studies, the diffused regions are removed by a short etching step in order to avoid increased recombination in the emitter and at the emitter surface.

With regard to the defect generation discussed in Chapter 4, Bothe *et al.* reported that phosphorus diffusions using optimized temperature ramps were capable of reducing the effective defect concentration in the degraded state by a factor of up to 3.5 [11]. The defect generation rate constant R_{gen}, however, was not affected by this treatment. The same is true for the defect annihilation discussed in Chapter 5. While the effective defect concentration after complete annihilation decreases after a P-diffusion with optimized temperature ramps, the annihilation rate constant R_{ann} does not change.

In order to investigate the impact of a P-diffusion on the permanent deactivation of the boron-oxygen complex, 1.4 Ω cm exclusively B-doped p-type Cz-Si is used. The interstitial oxygen concentration is $[O_i] = (7.5 \pm 0.5) \times 10^{17}$ cm^{-3}, as determined by means of Fourier transform infrared spectroscopy. The samples were damage-etched and RCA-cleaned before a 100 $\Omega/\square$ P-diffusion was performed at 847°C for half the samples. The resulting n^+ layers on both wafer sides were removed by acidic etching. After another RCA cleaning the sample surfaces were then passivated using PECVD silicon nitride on both sides of each sample.

Figure 6.2 shows the time dependence of the lifetime τ at a well-defined injection density of $\Delta n = 10^{15}$ cm^{-3} before (black circles) and after a P-diffusion step (red triangles). The samples are held at 135°C under illumination at 70 mW/cm^2. During the first 30 minutes, the lifetime decreases due to the formation of the boron-oxygen-related recombination center (which is accelerated by the increased temperature). However, after 30 minutes the lifetime begins to recover. After 140 h, τ saturates, at which point the samples are removed from the hot plate (as indicated by the dashed line). The samples are then placed under illumination at 10 mW/cm^2 and 25°C for more than 150 h. During that time the lifetime is constant within the uncertainty range of the QSSPC measurement, which is approximately ±5%, as can be

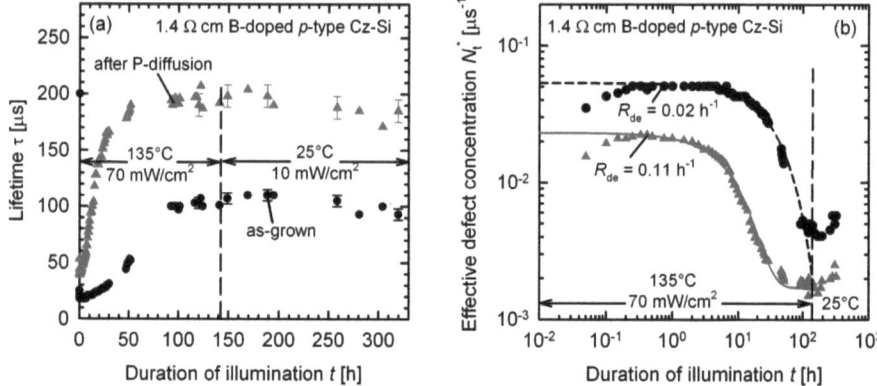

Figure 6.2: (a) Time dependence of the lifetime τ of a 1.4 Ω cm B-doped p-type Cz-Si sample before (black circles) and after P-diffusion (red triangles) under illumination at 135°C and at a light intensity of 70 mW/cm^2 (up to the dashed line). After 140 h, the samples are removed from the hot plate and illuminated at 25°C and 10 mW/cm^2 light intensity. No light-induced degradation is observed over the course of more than 150 h. (b) Effective defect concentration N_t^* derived from the lifetime data shown in Fig. 6.2(a), plotted on a double logarithmic scale. The solid lines represent fits of an exponential decay function, which yield the rate constant R_{de} of the deactivation process.

seen from Fig. 6.2(a).

In order to determine the rate constant of the deactivation process, the lifetimes from Fig. 6.2(a) are converted to effective defect concentrations N_t^*, as is shown in Fig. 6.2(b). The data can then be fitted by an exponential function of the form $y = y_0 + a\exp(-R_{de}t)$, where R_{de} is the deactivation rate constant. By this, $R_{de} = 0.11$ h^{-1} is obtained for the P-diffused sample, whereas $R_{de} = 0.02$ h^{-1} is determined for the as-grown sample. The phosphorus diffusion process thus increases the deactivation rate constant considerably. In addition, the final effective defect concentration after deactivation in the P-diffused sample is lower than in the as-grown sample.

Performing deactivation experiments of the type shown in Fig. 6.2 at different temperatures in the range between 135°C and 215°C reveals that the deactivation process is thermally activated. Figure 6.3 shows the deactivation rate constant R_{de} obtained from the exponential fits in Fig. 6.2 as a function of inverse temperature $1000/T$. Up to the maximum temperature of 215°C the data can be fitted by an Arrhenius law

$$R_{de} = \nu \exp\left(-E_{de}/k_B T\right), \tag{6.1}$$

where ν is the attempt frequency and E_{de} is the activation energy of the permanent deactivation

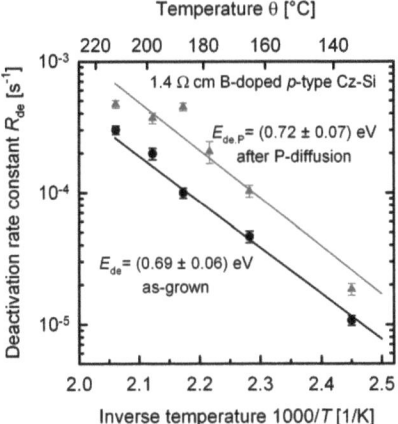

Figure 6.3: Arrhenius plot of the deactivation rate constant R_{de} in 1.4 Ω cm B-doped Cz-Si before (black circles) and after a P-diffusion step (red triangles). The illumination intensity during the deactivation was set at 70 mW/cm². The Arrhenius fits give activation energies of $E_{de} = (0.69 \pm 0.06)$ eV and $E_{de.P} = (0.72 \pm 0.07)$ eV before and after P-diffusion, respectively.

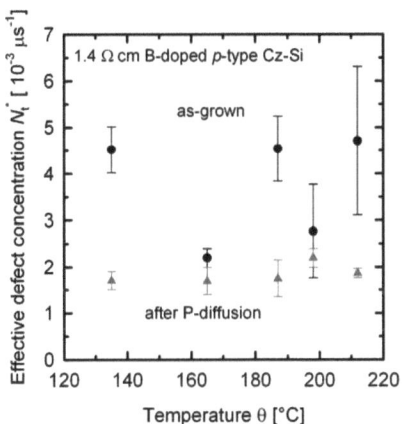

Figure 6.4: Effective defect concentration N_t^* in 1.4 Ω cm B-doped Cz-Si after deactivation of the BO complex as a function of annealing temperature. After phosphorus diffusion (red triangles), the defect concentration after deactivation is noticeably lower and shows less scattering than without phosphorus diffusion (black circles).

process.

After a P-diffusion (red triangles), the fit in Fig. 6.3 yields $E_{de.P} = (0.72 \pm 0.07)$ eV and $\nu_P = (2.0 \pm 0.5) \times 10^4$ s^{-1}. For the as-grown Cz-Si samples (black circles) a similar E_{de} of

(0.69 ± 0.06) eV is obtained, however, the attempt frequency is one order of magnitude lower: $\nu = (4 \pm 2) \times 10^3$ s^{-1}.

The final effective defect concentration N_t^* after permanent deactivation as a function of annealing temperature is plotted in Fig. 6.4. No clear correlation between the temperature and N_t^* is observable. Instead, the data scatters quite strongly, especially for the as-grown samples. However, the effective defect concentration in the samples that underwent a P-diffusion is lower than in the as-grown samples. In addition, the scatter of the N_t^* data is much weaker after P-diffusion.

6.1.2 Impact of silicon nitride deposition

Silicon nitride (SiN$_x$) is commonly used in crystalline silicon solar cell production as an antireflective coating on the front side of the cell. The standard deposition technique is plasma-enhanced chemical vapor deposition (PECVD) at temperatures between 400°C and 500°C. Apart from its optical qualities, the SiN$_x$ coating is also known to have beneficial effects on the bulk lifetime of multicrystalline silicon due to the so-called hydrogen-passivation: PECVD-SiN$_x$ contains high amounts of hydrogen (H at.% ∼16-18% [44, 45]). During firing of the metal contacts at around 850°C for a couple of seconds, the hydrogen in the SiN$_x$ becomes very mobile and partly diffuses into the silicon bulk, where it may then passivate defects.

With regard to boron-oxygen-related defect generation and annihilation, no impact of SiN$_x$ deposition has been observed so far. In contrast, Münzer [46] recently reported that in his experiments defect deactivation could only be observed on solar cells which were coated with hydrogen-rich SiN$_x$ films.

In this work, the impact of depositing a hydrogen-rich silicon nitride layer on the permanent deactivation process is investigated via lifetime measurements for the first time. The material used is 0.9 Ω cm B-doped Cz-Si which is passivated with either PECVD-SiN$_x$ or plasma-assisted atomic layer deposited (PA-ALD) aluminum oxide (Al$_2$O$_3$). PA-ALD-Al$_2$O$_3$ contains very little hydrogen (H at.% ∼3% [47]) and has recently been found to be extremely effective in passivating the surface of p-type silicon samples [48–50].

Sample preparation included acidic damage etching, RCA cleaning and a phosphorus diffusion. The n^+ layers on both sides of the wafer were removed by acidic etching before the samples were RCA-cleaned again. Subsequently, the samples were split into three groups: (1) surface passivation with PECVD-SiN$_x$ using an Oxford Instruments Plasmalab 80 Plus reactor. This static setup uses a remote plasma excited in a cavity outside the deposition chamber. The used process gases are silane (SiH$_4$), ammonia (NH$_3$) and nitrogen (N$_2$). (2) Surface passivation with PECVD-SiN$_x$ using a Roth & Rau SiNA system. The SiNA is an inline deposition system which works with a remote plasma that is excited by a linear microwave antenna which is located directly above the silicon wafers. In contrast to the static reactor, the samples are thus exposed to the plasma during deposition in the SiNA system, which may result in some

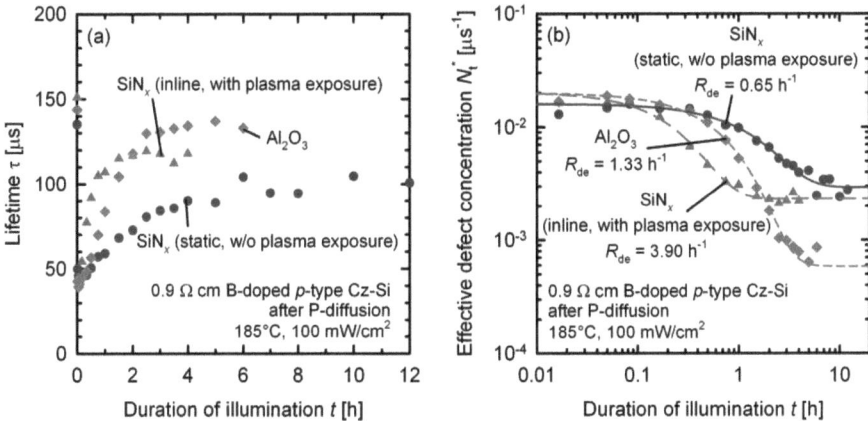

Figure 6.5: (a) Lifetime τ as a function of duration of illumination t at 185°C and 100 mW/cm² light intensity (halogen lamp) of three 0.9 Ω cm B-doped p-type Cz-Si samples. Two of the samples are passivated with PECVD-SiN$_x$, deposited in two different types of reactors, while the third sample is passivated with ALD-Al$_2$O$_3$. (b) Effective defect concentration N_t^*, derived from the lifetime data in Fig. 6.5(a), as a function of duration of illumination t, plotted on a double logarithmic scale. The data are fitted by an exponential function (indicated by the lines), which yields the deactivation rate constant R_{de}.

plasma damage at the surface of the wafers. The process gases are SiH$_4$, NH$_3$, N$_2$ and hydrogen (H$_2$). (3) Surface passivation with plasma-assisted ALD-Al$_2$O$_3$ using an Oxford Instruments FlexAL™ reactor. Atomic layer deposition is done with a remote plasma excited by inductive coupling in a cavity and trimethylaluminum (Al$_2$(CH$_3$)$_6$) and oxygen (O$_2$) as process gases.

The sample temperature during SiN$_x$ deposition is 400°C. The thickness of the SiN$_x$ layer is 70 nm with a refractive index of 2.4 (at a wavelength of λ = 632 nm). During atomic layer deposition, the substrate temperature is set at 200°C. The Al$_2$O$_3$ layer has a thickness of 30 nm and subsequent to the deposition the samples are annealed for 15 minutes at 425°C to activate the surface passivation [48].

Figure 6.5 depicts the permanent deactivation of the BO complex in three 0.9 Ω cm B-doped p-type Cz-Si samples with one of the three passivations each. Figure 6.5(a) shows the lifetime τ, extracted at a fixed injection level of $\Delta n = 1.8 \times 10^{15}$ cm^{-3}, as a function of time t under illumination at 185°C and 100 mW/cm² (halogen lamp). Before $t = 0$, the BO center had been fully annihilated by a 10-minutes anneal at 200°C in darkness, resulting in comparable lifetimes of 130 to 150 μs for all three samples. Subsequently, the lifetime drastically decreases, due to the accelerated generation of BO centers at this elevated temperature. After 5 minutes, however, the lifetime begins to increase again until it saturates after a few hours.

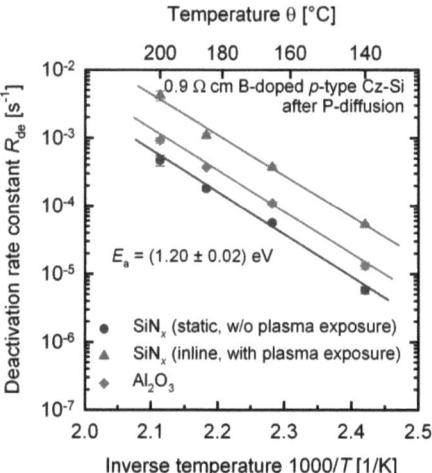

Figure 6.6: Arrhenius plot of the deactivation rate constants R_{de} determined in the three groups of samples. The activation energy $E_a = (1.20 \pm 0.02)$ eV is the same for all samples. The deactivation rate constants measured for the SiN$_x$-coated samples with plasma exposure (red triangles), however, are one order of magnitude larger than those measured in the samples without plasma exposure (blue circles).

In Fig. 6.5(a), the difference between the three passivation techniques is clearly discernible: the lifetime of the sample that was passivated using the SiNA inline system with plasma exposure (red triangles) saturates after 2 hours, whereas the lifetime of the sample that was passivated using the static Plasmalab 80 Plus reactor without plasma exposure (blue circles) saturates after 10 hours.

This difference becomes even more obvious in the deactivation rate constants R_{de}, which are shown in Fig. 6.5(b). Here, $R_{de} = 0.65$ h^{-1} is obtained for the sample that was not exposed to the plasma during PECVD-SiN$_x$ deposition (blue circles), while $R_{de} = 3.90$ h^{-1} is obtained for the sample that was exposed to plasma during PECVD-SiN$_x$ deposition (red triangles). As a key result of this Section, exposure to a plasma during deposition thus seems to increase the deactivation rate constant by almost one order of magnitude. In contrast, the absolute hydrogen content of the passivating dielectric layer does not have an impact on the deactivation process by itself, as is demonstrated by the fact that the deactivation rate constant R_{de} of the Al$_2$O$_3$ passivated sample is 1.33 h^{-1} (grey diamonds), which is only slightly higher than that of the SiN$_x$-coated sample not exposed to plasma.

In addition to the experiments shown in Fig. 6.5, the deactivation rate constants R_{de} are also determined at 140°C, 165°C and 200°C. In Fig. 6.6, these deactivation rate constants are plotted versus the inverse temperature $1000/T$ in an Arrhenius plot. For all investigated temperatures, the deactivation rate constant of samples that were exposed to plasma during SiN$_x$ deposition is one order of magnitude higher than that of the samples not exposed to plasma, confirming the result presented in Fig. 6.5. The activation energy $E_\text{a} = (1.20 \pm 0.02)$ eV is the same for all samples.

This new finding can also consistently explain the results presented in Ref. [46], where solar cells coated with almost hydrogen-free silicon nitride deposited via low-pressure chemical vapor deposition (LPCVD) were compared to solar cells coated with hydrogen-rich silicon nitride deposited via PECVD. Since deactivation could only be observed in the solar cells that were coated with hydrogen-rich PECVD-SiN$_x$, it was concluded that hydrogen was vital for the deactivation process. However, in light of the new findings presented in this work, it seems more likely that the different deposition techniques are the reason for the different deactivation behavior, since LPCVD, in contrast to the deposition techniques used in this work (i.e., PECVD and PA-ALD), is not a plasma process.

This conclusion in turn suggests that a precursor for the deactivation process is created in the presence of a plasma. Exposure to the plasma during deposition could then either increase the amount of this precursor or facilitate the in-diffusion of the precursor into the silicon bulk. A possible candidate for this precursor is some form of hydrogen. Hydrogen is present both during PA-ALD of Al$_2$O$_3$ and PECVD of SiN$_x$ and it is known that surface damage caused by a PECVD passivation process can increase in-diffusion of hydrogen into silicon [51]. In addition, it was recently reported that the amount of hydrogen that diffuses from a SiN$_x$:H layer into the silicon bulk is sensitive to the deposition conditions [52].

Note that the activation energy of 1.2 eV is notably higher than the 0.7 eV presented in the previous Section. This deviation can be attributed to the different resistivities of the investigated samples (1.4 Ω cm in Sec. 6.1.1 and 0.9 Ω cm in this Section, respectively), as will explained in the following Section.

6.1.3 Impact of boron concentration

In Chapter 4 and 5, is was shown that both the defect generation and the defect annihilation depend on the hole concentration p_0. The defect concentration N_t^* after complete degradation is proportional to p_0, while the defect generation rate constant R_gen is proportional to the square of the doping concentration p_0^2. The defect annihilation rate constant R_ann, on the other hand, is inversely proportional to p_0.

In order to investigate the dependence of the permanent deactivation process on the boron concentration N_A, three samples with varying N_A but similar interstitial oxygen concentrations are studied. The resistivities ρ of the samples are 3.3 Ω cm, 1.08 Ω cm and 0.43 Ω cm, respectively. In order to remove fast-diffusing metal impurities, the samples underwent a phosphorus diffusion. The resulting n^+ layers on both sides of the wafer were subsequently removed and the sample surfaces were passivated using ALD-Al$_2$O$_3$. The permanent deactivation is done at

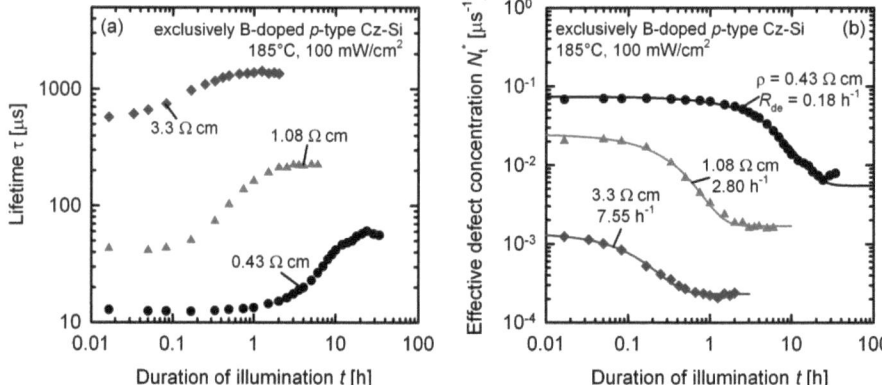

Figure 6.7: (a) Lifetime τ of three exclusively B-doped p-type Cz-Si samples with different resistivities plotted versus time t at 185°C under illumination at 100 mW/cm² on a double logarithmic scale. (b) Effective defect concentration N_t^* derived from the lifetime data shown in Fig. 6.7(a), plotted versus time t at 185°C under illumination on a double logarithmic scale. The solid lines represent fits of an exponential decay function that yield the deactivation rate constant R_{de}.

185°C and at a light intensity of 100 mW/cm². The lifetime is extracted at a fixed injection level of $\Delta n = 0.1 \times p_0$.

The time dependence of the lifetime τ in the three samples is shown in Fig. 6.7(a) on a double logarithmic scale. At time $t = 0$, the defect is fully annihilated (not shown in Fig. 6.7). During the first minutes, the lifetime decreases due to the very fast defect generation at 185°C. Subsequently, however, the lifetime increases again. Depending on the resistivity, the lifetime saturates after 1 hour (3.3 Ω cm sample, blue diamonds), 3 hours (1.08 Ω cm sample, red triangles) or 20 hours (0.43 Ω cm sample, black circles). In addition, the boron concentration has a clear impact on the level of the recovered lifetime. In the 3.3 Ω cm sample, τ saturates at 1380 µs, while in the 0.43 Ω cm wafer, τ saturates at 58 µs.

In order to determine the deactivation rate constant R_{de}, the lifetime data from Fig. 6.7(a) is converted to the effective defect concentration N_t^*, as shown in Fig. 6.7(b). As can be seen, the deactivation rate constant R_{de} decreases significantly with decreasing resistivity ρ, i.e. with increasing boron concentration N_A. The 3.3 Ω cm sample (blue diamonds) has a deactivation rate constant of 7.55 h^{-1}, while the 1.08 Ω cm sample (red triangles) has a deactivation rate constant of 2.80 h^{-1} and the 0.43 Ω cm sample (black circles) has a deactivation rate constant of 0.18 h^{-1}.

The deactivation rate constants determined from Fig. 6.7(b) are plotted versus the boron concentration N_A in Fig. 6.8 on a double logarithmic scale. The data can be described by a

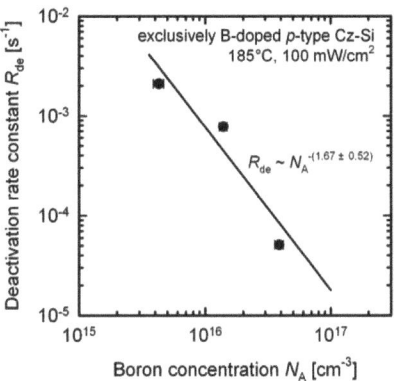

Figure 6.8: Deactivation rate constants R_{de} determined under illumination at 185°C and 100 mW/cm² light intensity in exclusively B-doped p-type Cz-Si plotted versus the boron concentration N_A on a double logarithmic scale. The fit with a power law (solid line) yields an exponent of $-(1.67 \pm 0.52)$, i.e. $R_{de} \propto N_A^{-(1.67 \pm 0.52)}$.

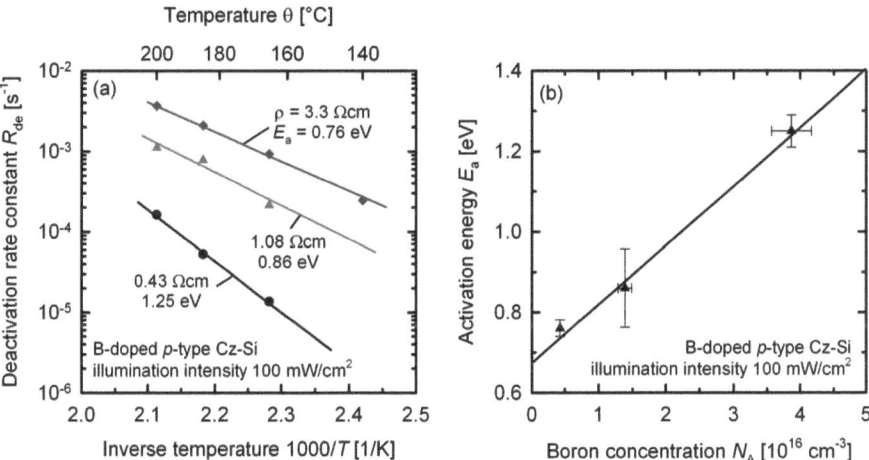

Figure 6.9: (a) Arrhenius plot of the deactivation rate constants R_{de} determined for three different resistivities ρ in exclusively B-doped p-type Cz-Si under illumination at 100 mW/cm². (b) Activation energies E_a determined in Fig. 6.9(a), plotted versus the boron concentration N_A.

power law with an exponent of $-(1.67 \pm 0.52)$, i.e. $R_{de} \propto N_A^{-(1.67 \pm 0.52)}$. It should be noted, however, that this data set is quite small and hence the fit has a high uncertainty.

Similar samples are also illuminated at 140°C, 165°C and 200°C, respectively. The measured deactivation rate constants R_{de} are plotted in Fig. 6.9(a) in an Arrhenius plot versus the inverse temperature $1000/T$. At all investigated temperatures, R_{de} increases with decreasing N_A. Interestingly, however, the difference in the R_{de} values determined for different boron concentrations increases with decreasing temperature. As a consequence, the permanent deactivation treatment at 140°C could not be finished for the 1.08 Ω cm sample (red triangles) and the 0.43 Ω cm sample (black circles).

The changing ratio of the deactivation rate constants at different temperatures is a result of the different activation energies E_a of the deactivation process in the different samples: an activation energy of 1.25 eV is found for the 0.43 Ω cm sample, whereas $E_a = 0.86$ eV is measured for the 1.08 Ω cm sample and $E_a = 0.76$ eV is obtained for the 3.3 Ω cm sample. The activation energy E_a thus increases with increasing boron concentration.

Plotting E_a versus N_A, as shown in Fig. 6.9(b), a linear dependence of the activation energy on the boron concentration is revealed. Again it should be noted that this data set is quite small and that a broader range of samples would be needed to verify this finding. However, looking at the activation energies obtained in the previous Sections, i.e. 1.2 eV on 0.9 Ω cm Cz-Si in Sec. 6.1.2 and 0.7 eV on 1.4 Ω cm Cz-Si in Sec. 6.1.1, a dependence of E_a on N_A seems evident.

6.1.4 Impact of interstitial oxygen concentration

In order to study the impact of the interstitial oxygen concentration [O_i] on the deactivation process, 0.72 Ω cm boron-doped Cz-Si wafers with varying [O_i] are examined. The interstitial oxygen concentrations in the samples are in the range of [O_i] = $(2.5 - 5.1) \times 10^{17}$ cm^{-3}, as determined according to DIN 50438-1 using a Bruker Equinox 55 Fourier transform infrared (FTIR) spectrometer.

The samples underwent a thermal donor annihilation step (i.e. annealing at 750°C for 30 minutes in N_2) and were RCA-cleaned before being passivated by either PECVD-SiN$_x$ (without exposure to the plasma during deposition) or by plasma-assisted atomic layer deposited aluminum oxide (PA-ALD-Al$_2$O$_3$).

In Fig. 6.10(a), the lifetime τ (extracted at a fixed injection level of $\Delta n = 10^{15}$ cm^{-3}) is plotted versus the duration of illumination t for three samples with different interstitial oxygen concentrations [O_i]. The samples are illuminated at 185°C with a halogen lamp of 100 mW/cm^2 light intensity. The saturation value of the recovered lifetime decreases significantly with increasing interstitial oxygen concentration: the sample with [O_i] = 2.67×10^{17} cm^{-3} recovers to a remarkably high lifetime of (690 ± 10) μs while the lifetime of the sample with [O_i] = 4.98×10^{17} cm^{-3} saturates at (410 ± 10) μs.

Figure 6.10(b) shows the measured effective defect concentration N_t^* (with $N_{t,\text{max}}^*$ normalized to 1), calculated from the lifetime data shown in Fig. 6.10(a), as a function of illumination time t. On the double logarithmic scale, the difference in the deactivation rate constants is easily

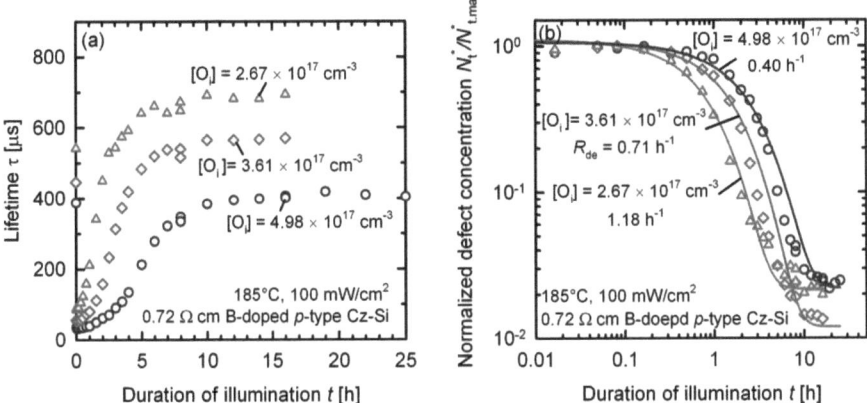

Figure 6.10: (a) Lifetime τ of three 0.72 Ω cm B-doped p-type Cz-Si samples plotted versus the duration of illumination t at 185°C and 100 mW/cm² light intensity. The interstitial oxygen concentrations [O_i] of the samples are 2.67×10^{17} cm^{-3} (triangles), 3.61×10^{17} cm^{-3} (diamonds), and 4.98×10^{17} cm^{-3} (circles), respectively. The saturation value of τ decreases with increasing [O_i]. (b) Normalized defect concentration $N_t^*/N_{t,\mathrm{max}}^*$ calculated from the lifetime data shown in Fig. 6.10(a) plotted on a double logarithmic scale versus the duration of illumination t at 185°C. The deactivation rate constant R_{de} is determined by an exponential fit to the data (solid lines).

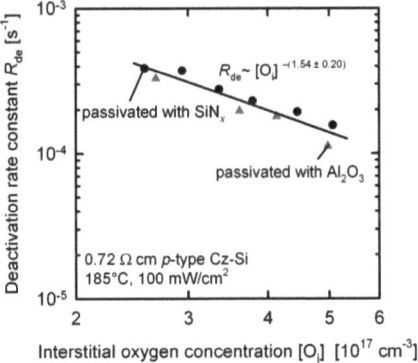

Figure 6.11: Measured deactivation rate constants R_{de} of 0.72 Ω cm B-doped p-type Cz-Si samples at 185°C and a light intensity of 100 mW/cm² plotted versus the interstitial oxygen concentration [O_i]. The circles correspond to samples passivated with PECVD-SiN$_x$ while the triangles correspond to samples passivated with PA-ALD-Al$_2$O$_3$. In accordance with Sec. 6.1.2, there is no discernible difference between the two sets of samples. The overall dependence can be described by a power law $R_{\mathrm{de}} \propto [O_i]^{-(1.54 \pm 0.20)}$ as indicated by the solid line.

distinguishable and it can be seen that the deactivation slows down with increasing interstitial oxygen concentration.

The deactivation rate constants R_{de} for all investigated samples are plotted versus the interstitial oxygen concentrations $[O_i]$ in Fig. 6.11. As can be seen from Fig. 6.11, the deactivation procedure takes longer the higher the interstitial oxygen content is. The linear dependence displayed on the double logarithmic scale reveals a power law with an exponent of $-(1.54 \pm 0.20)$, i.e. $R_{de} \propto [O_i]^{-(1.54 \pm 0.20)}$.

Figure 6.11 includes both the samples passivated with PECVD-SiN$_x$ (black circles) and the samples passivated with PA-ALD-Al$_2$O$_3$ (red triangles). The deactivation of the boron-oxygen-related centers proceeds in the same way in all samples, confirming the results presented in Sec. 6.1.2.

6.1.5 Impact of thermal donors

Thermal donors (TD) are oxygen clusters generated during long-term annealing at 450°C [53, 54]. Apart from affecting the free carrier concentration through compensation, high concentrations of thermal donors were also found to reduce the concentration of the boron-oxygen-related recombination-center by more than a factor of 3 [11].

In order to study the impact of thermal donors on the deactivation of the boron-oxygen complex, 0.8 Ω cm B-doped Cz-Si with an interstitial oxygen concentration of $[O_i] = (7.5 \pm 0.5) \times 10^{17}$ cm^{-3} is studied. All samples underwent acidic damage etching and thermal donor annihilation, followed by RCA cleaning and PECVD-SiN$_x$ coating. The SiN$_x$ coating was used as a diffusion barrier to prevent metal contamination during the long-term anneals at 450°C. Note that all samples were coated with SiN$_x$, even those that were not annealed at 450°C, to ensure an as similar processing sequence as possible for all samples.

The samples were then split into four groups: the control samples and three groups that were annealed at 450°C in air for 8 hours, 16 hours and 32 hours, respectively. Subsequently, the silicon nitride was removed from all samples by a short etch in HF, the samples were RCA-cleaned and passivated by plasma-assisted ALD-Al$_2$O$_3$. The atomic layer deposition was followed by a 15 min anneal at 425°C to activate the surface passivation. The thermal donor concentration was determined by measuring the resistivity of the samples before and after annealing.

The treatment at 450°C results in thermal donor concentrations of $[TD]_{8h} = 1.7 \times 10^{15}$ cm^{-3}, $[TD]_{16h} = 2.2 \times 10^{15}$ cm^{-3} and $[TD]_{32h} = 4.0 \times 10^{15}$ cm^{-3}, with a respective uncertainty of $[TD]_{err} = \pm 0.3 \times 10^{15}$ cm^{-3}. No changes in resistivity exceeding this uncertainty are observed either after annealing the Al$_2$O$_3$ layers (15 min at 425°C) or after the deactivation procedure (up to 40 hours at 185°C).

Figure 6.12(a) depicts the time dependence of the carrier lifetime τ (at an injection level of $\Delta n = 2 \times 10^{15}$ cm^{-3}) during illumination at 185°C for all samples. The control wafer (red

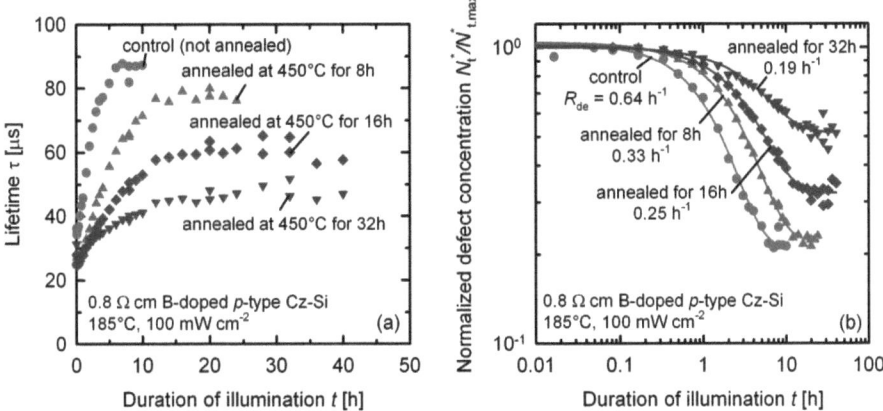

Figure 6.12: (a) Time dependence of the carrier lifetime τ during illumination at 185°C and 100 mW/cm^2 light intensity of four B-doped p-type Cz-Si samples. Three of the samples were annealed at 450°C for 8 h, 16 h, and 32 h, respectively. The saturation value of the lifetime after defect deactivation decreases with increasing annealing time. (b) Normalized defect concentration $N_t^*/N_{t,\text{max}}^*$ derived from the lifetime data in Fig. 6.12(a) plotted versus the duration of illumination t on a double logarithmic scale. The deactivation rate constant R_{de} decreases with increasing annealing time, from 0.64 h^{-1} (non-annealed control, red circles) to 0.19 h^{-1} (annealed for 32 h at 450°C, purple triangles down). In addition, the deactivation process is less effective after thermal donor formation.

circles), without any thermal donors, displays the fastest deactivation as well as the highest recovered lifetime of (87 ± 5) µs. With increasing thermal donor concentration, the deactivation slows down and the saturation value for the recovered lifetime decreases, leading to a final value of (77 ± 3) µs for the sample which had been annealed at 450°C for 8 h (green triangles up), (60 ± 5) µs for the sample annealed for 16 h (blue diamonds) and (45 ± 5) µs for the sample which was annealed for 32 h (purple triangles down).

Figure 6.12(b) shows the effective defect concentration N_t^* (normalized to $N_{t,\text{max}}^*$), derived from the lifetimes shown in Fig. 6.12(a), as a function of the duration of illumination t on a double logarithmic scale. Note that the initial lifetimes measured before light-induced degradation τ_0 are not shown in Fig. 6.12(a) since they are above 100 µs (i.e., 120 to 200 µs). The control wafer without any thermal donors yields a deactivation rate constant of $R_{\text{de}} = 0.64$ h^{-1}, which is more than 3 times the deactivation rate constant of the sample which was annealed at 450°C for 32 h ($R_{\text{de}} = 0.19$ h^{-1}). Figure 6.12(b) also shows that the final normalized defect concentration increases with increasing TD concentration. Apart from slowing down the deactivation process, thermal donors thus also seem to reduce its efficacy.

Since thermal donors are oxygen clusters, their impact on the deactivation rate constant and deactivation extent may be further indirect evidence that oxygen-related defects are involved in the deactivation process. However, as will be discussed in Chapter 7, the reduction of the deactivation rate constant could also be related to other processes occurring during prolonged annealing at 450°C.

6.1.6 Impact of compensation

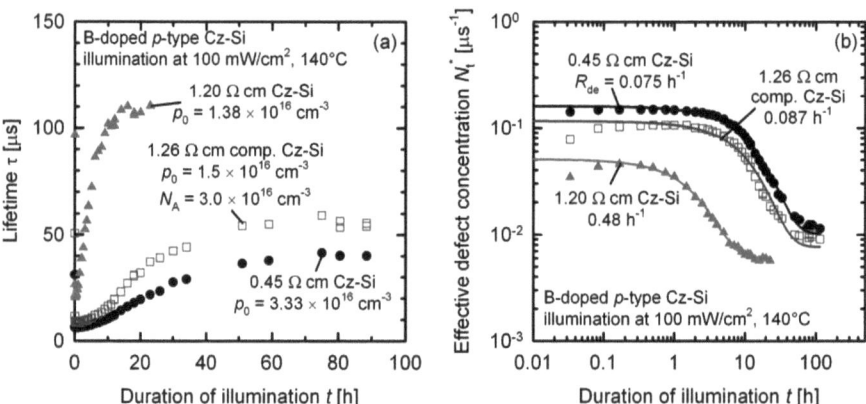

Figure 6.13: (a) Lifetime τ plotted versus time t at 140°C under illumination at 100 mW/cm^2 of three p-type Cz-Si samples, two of which are exclusively doped with boron (red triangles and black circles) while one is doped with both boron and phosphorus (open purple squares). (b) Effective defect concentration N_t^* derived from the lifetime data shown in Fig. 6.13(a), plotted versus time t on a double logarithmic scale. The deactivation rate constants R_{de} are determined through exponential fits of the data (solid lines).

In Section 6.1.3, it was shown that the rate constant of the deactivation decreases with increasing boron concentration. In this Section, the impact of compensation on the deactivation process is investigated. Again, an interesting aspect of compensated p-type silicon is the difference between the (substitutional) boron concentration N_A and the net doping concentration p_0 (as opposed to exclusively B-doped silicon, where $N_A = p_0$).

For this investigation, two sets of samples (similar to the ones used in Sec. 5.2.1) are studied: the first set consists of two wafers that were cut from two different ingots (44 and 45), whereas the second set of samples consists of six wafers that were all cut from the same ingot (Ingot A). For sample details see Sec. 5.2.1 and Tab. 3.2. Sample preparation included a P-diffusion, removal of the resulting n^+ layers on both sides of the wafer and surface passivation using PECVD-SiN$_x$.

Figure 6.13 depicts the deactivation of the BO complex in two non-compensated B-doped

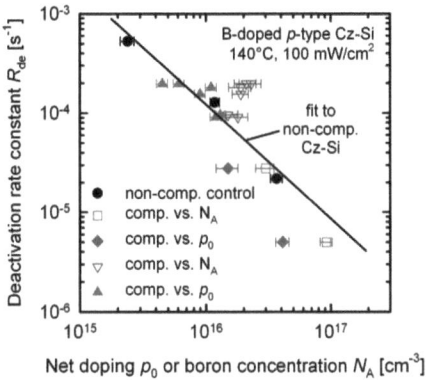

Figure 6.14: Measured deactivation rate constants R_{de} of dopant-compensated p-type Cz-Si samples at 140°C and a light intensity of 100 mW/cm^2 plotted on a double logarithmic scale versus the net doping concentration p_0 (filled symbols) and the boron concentration N_A (open symbols), respectively. Also plotted are R_{de} values obtained on non-compensated control samples (black circles). No unambiguous dependence on either p_0 or N_A is discernible.

Cz-Si samples and a B- and P-doped compensated p-type sample (#45). Figure 6.13(a) shows the lifetime τ as a function of time t at which the samples are illuminated at 140°C and a light intensity of 100 mW/cm^2. Looking at the lifetime data, which is extracted at a fixed injection level of $\Delta n = 0.1 \times p_0$, it is already obvious that the deactivation proceeds much faster in the non-compensated 1.20 Ω cm sample (τ saturates after 20 hours) than in the compensated 1.26 Ω cm sample (τ saturates after about 80 hours).

This difference becomes even more evident when looking at the deactivation rate constants R_{de} determined in Fig. 6.13(b): the effective defect concentration N_t^* is plotted as a function of time t and an exponential decay function is fitted to the data. By this, $R_{de} = 0.48$ h^{-1} is obtained for the non-compensated 1.20 Ω cm sample, while $R_{de} = 0.087$ h^{-1} is determined for the compensated 1.26 Ω cm sample. Interestingly, R_{de} equals 0.075 h^{-1} for the non-compensated 0.45 Ω cm sample, which is comparable to the deactivation rate constant of the compensated sample.

Looking at the net doping concentrations p_0 and the boron concentrations N_A of the three samples, it thus seems that the deactivation rate constant correlates with N_A (which is similar in the non-compensated 0.45 Ω cm sample and in the compensated 1.26 Ω cm sample).

However, looking at the deactivation rate constants determined for all compensated samples, the result is ambiguous, as is shown in Fig. 6.14. As a reference, data from non-compensated B-doped Cz-Si samples (black circles) is included and fitted by a power law (solid line). The deactivation rate constants R_{de} of samples 44 and 45 are plotted versus the net doping concentration p_0 (filled green diamonds) as well as versus the boron concentration N_A (open pink squares). When compared to the control data of the non-compensated samples (black circles), the R_{de} values of the compensated samples agree better when plotted versus the boron concentration N_A. On the other hand, when looking at the six samples that were cut from Ingot A, the data agrees better when R_{de} is plotted versus p_0 (filled red triangles up) than when it is plotted versus N_A (open blue triangles down).

6.1.7 Potential of the carrier lifetime after permanent deactivation of the BO defect

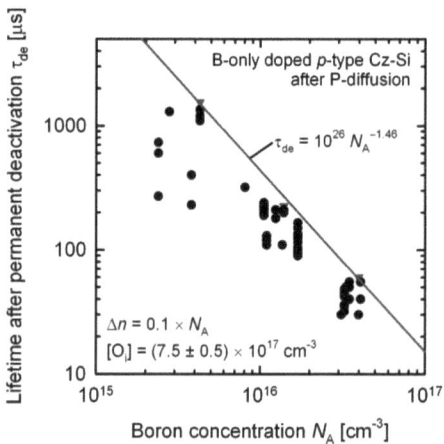

Figure 6.15: Carrier lifetimes τ_{de} after permanent deactivation of the BO defect as a function of the boron concentration N_A, plotted on a double logarithmic scale. The depicted samples are exclusively doped with boron and underwent a phosphorus diffusion step. The blue triangles correspond to a single set of samples in which the average lifetime after permanent deactivation was highest. The solid blue line indicates a fit to that data, which yields a power law $\tau_{de} = 10^{26} \, N_A^{-1.46}$ (τ_{de} in μs and N_A in cm^{-3}) to describe the potential of the carrier lifetime after permanent deactivation of the BO defect.

In the previous Sections, it was shown that the kinetics of the permanent deactivation process depend on a variety of parameters. In addition, the lifetime after permanent deactivation was found to vary strongly, even in samples that were cut from the same wafer and underwent the same processing steps. A summary of carrier lifetimes τ_{de} measured after permanent deactivation of the boron-oxygen defect, plotted as a function of the boron concentration N_A on a double logarithmic scale, is shown in Fig. 6.15. Since a phosphorus diffusion clearly improved the lifetime after permanent deactivation and since most solar cell process sequences include a

P-diffusion, Fig. 6.15 only depicts samples that underwent a P-diffusion step. In addition, the interstitial oxygen concentration in all samples is $[O_i] = (7.5 \pm 0.5) \times 10^{17}$ cm^{-3}.

Not surprisingly, an overall trend of decreasing lifetime τ_{de} with increasing N_A is observed. However, when comparing lifetimes measured in samples with the same doping concentration, variations by up to a factor of 2.5 are observed. The cause for these variations has not been identified so far.

A parameterization of the experimental data shown in Fig. 6.15 is not feasible, however, it can be used to assess the potential of the carrier lifetime after permanent deactivation of the boron-oxygen defect. The blue triangles in Fig. 6.15 correspond to a single set of samples, in which the highest lifetimes after permanent deactivation were measured. A fit to these data points (indicated by the solid line) yields a power law of the form $\tau_{de} = 10^{26} N_A^{-1.46}$ (τ_{de} in μs and N_A in cm^{-3}), which will be used in Section 8.2 to estimate the efficiency potential of screen-printed solar cells fabricated on B-doped Cz-Si after permanent deactivation of the BO defect.

6.2 Stability of the deactivated state

The bulk lifetime (see Sec. 6.1.1), the open-circuit voltage, and the efficiency of solar cells (Sec. 8.1 and [20]) have been demonstrated to be stable under illumination at 25°C for more than 100 hours after applying the permanent deactivation treatment. However, under certain conditions a degradation of the bulk lifetime after application of the deactivation procedure is observed.

In this Section, the destabilization of the deactivated state of the boron-oxygen defect is investigated under three different conditions: (1) illumination at room temperature, (2) long-term illumination at elevated temperature, and (3) extended annealing in darkness. With regard to a permanent improvement of solar cells made on low-resistivity boron-doped Cz-Si, any instability of the deactivated state is obviously unfavorable. With regard to understanding the underlying defect reaction, however, this renewed degradation could provide useful insights into the reaction kinetics.

6.2.1 Partial degradation under illumination at room temperature

Even though many samples exhibit perfectly stable lifetimes after application of the deactivation treatment (as shown in Fig. 6.2), partial degradation as depicted in Fig. 6.16 is also frequently observed. Figure 6.16(a) depicts the lifetime τ in a 0.72 Ω cm B-doped p-type Cz-Si sample doped exclusively with boron during the permanent deactivation treatment and subsequently during illumination at room temperature. The lifetime is measured using the QSSPC technique and is extracted at a fixed injection level of $\Delta n = 10^{15}$ cm^{-3}.

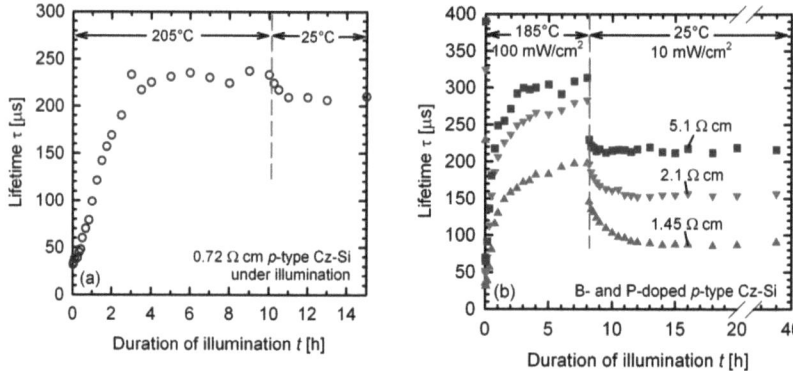

Figure 6.16: (a) Partial degradation of the carrier lifetime τ in B-only doped p-type Cz-Si under illumination at room temperature after applying the deactivation treatment at 205°C and a light-intensity of 100 mW/cm². (b) Partial degradation of the carrier lifetime τ in dopant-compensated p-type Cz-Si under illumination at room temperature after applying the deactivation treatment at 185°C and a light-intensity of 100 mW/cm².

At the beginning of the experiment, the sample is illuminated at 205°C and at a light intensity of 100 mW/cm² for 10 hours (up to the dashed red line). Subsequently, the sample is illuminated at 25°C and 10 mW/cm² light intensity. During illumination at 205°C, the lifetime increases from 30 μs (in the fully degraded state) to 235 μs due to the permanent deactivation of the boron-oxygen-related recombination center. During subsequent illumination at room temperature, the lifetime again decreases, however, the saturation value of 210 μs that is reached after 3 hours is still significantly higher than the initial 30 μs in the degraded state.

Figure 6.16(b) shows the partial degradation under illumination at room temperature after application of the permanent deactivation procedure in three B- and P-doped p-type Cz-Si samples with different resistivities ρ. The lifetime is measured using the QSSPC technique and is extracted at a fixed injection level of $\Delta n = 0.1 \times p_0$. Permanent deactivation is done at 185°C and a light intensity of 100 mW/cm². After 8 hours, the samples are removed from the hot plate (as indicated by the dashed red line) and are subsequently illuminated at 25°C and at a light intensity of 10 mW/cm².

In the 5.1 Ω cm sample (purple squares), the initial lifetime (in the degraded state) was 65 μs. During permanent deactivation of the BO complex, the lifetime increases to 310 μs. However, during subsequent illumination at room temperature, the lifetime again decreases and saturates at an intermediate level of 230 μs. Similarly, in the 1.45 Ω cm sample (green triangles up), the degraded lifetime is 31 μs, the lifetime after permanent deactivation is 198 μs and the lifetime after partial degradation during illumination at room temperature is 87 μs.

Note that the degrading lifetime saturates after 2 hours in the 5.1 Ω cm sample (purple squares) but after 10 hours in the 1.45 Ω cm sample (green triangles up). Interestingly, it thus seems as if the partial degradation after the deactivation treatment proceeds faster with decreasing doping concentration p_0. This is especially interesting since the light-induced degradation discussed in Chapter 4 was found to proceed slower with decreasing doping concentration.

6.2.2 Complete degradation during long-term illumination at elevated temperature

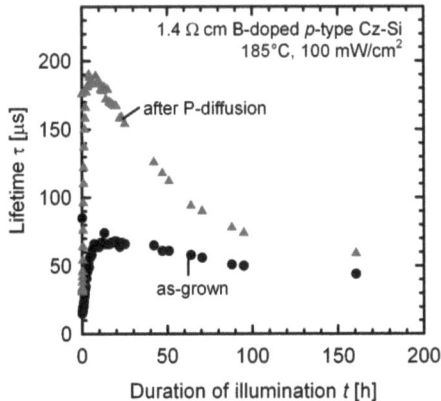

Figure 6.17: Complete degradation of the carrier lifetime τ in 1.4 Ω cm B-doped p-type Cz-Si under long-term illumination at elevated temperature after successful permanent deactivation of the BO defect. Shown are the lifetimes in an as-grown sample (black circles) and a sample that underwent a phosphorus diffusion step (red triangles). The samples are illuminated at 185°C and at a light-intensity of 100 mW/cm². Initially, this treatment results in a recovery of the lifetime due to the permanent deactivation of the BO defect, however, after 8 hours (after P-diffusion) and 20 hours (in the as-grown state), respectively, the lifetime starts to decrease again.

As the aim of defect deactivation is the recovery of lifetime, deactivation experiments are usually terminated, i.e. the samples are removed from illumination on the hot plate, once the increasing lifetime begins to saturate. As was pointed out before, the lifetime in the majority of samples was then found to be stable under illumination at room temperature, while in a couple of samples partial degradation under illumination at room temperature was observed.

In addition, prolonged illumination at elevated temperature was reported to result in complete degradation of the carrier lifetime after initially successful application of the permanent deactivation treatment [55]. In contrast to the aforementioned partial degradation, which was only observed in part of the samples examined in this work, prolonged illumination at elevated temperature always resulted in renewed degradation of the lifetime in the course of this work.

An example of this degradation phenomenon is shown in Fig. 6.17. The lifetime in two 1.4 Ω cm B-doped Cz-Si samples, one of which underwent a P-diffusion step (red triangles) and one of which did not (black circles), is depicted as a function of time t at which the samples are illuminated at 185°C and 100 mW/cm^2.

Similar to earlier results, the lifetime in the sample that underwent a P-diffusion (red triangles) begins to saturate after 4 hours, whereas the lifetime in the as-grown sample (black circles) begins to saturate after 8 hours. The lifetime is then stable for an additional 4 hours (after P-diffusion) and 12 hours (without P-diffusion), respectively. If illumination is further continued, the lifetime begins to decrease again. In the case shown, this renewed degradation continues for more than 150 hours (after which the experiment was terminated) and is likely to continue even further.

6.2.3 Complete degradation through extended annealing in darkness

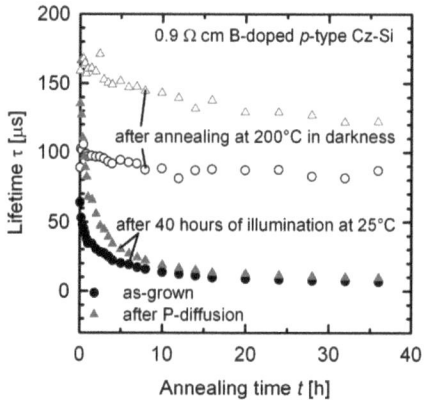

Figure 6.18: Complete degradation of the carrier lifetime τ in 0.9 Ω cm B-doped p-type Cz-Si after permanent deactivation of the BO defect. The degradation is caused by extended annealing at 200°C in darkness. The phenomenon is shown for an as-grown sample (black circles) and a sample that underwent a P-diffusion (red triangles). The open symbols correspond to lifetimes measured directly after annealing in darkness, whereas the filled symbols correspond to lifetimes measured after 40 hours of illumination at 25°C.

Complete degradation of the lifetime after applying the deactivation treatment was also reported after annealing in darkness [20]. An example of this effect is shown for two 0.9 Ω cm B-doped p-type Cz-Si samples, one that underwent a P-diffusion step (red triangles) and one in the as-grown state (black circles), in Fig. 6.18. Deactivation of the boron-oxygen-related recombination center is done at 200°C at a light intensity of 100 mW/cm^2. The lifetime begins to saturate after 30 minutes (after P-diffusion) and 2 hours (without P-diffusion), respectively, at which point the samples are removed from the hot plate.

In Fig. 6.18, the lifetime at $t = 0$ corresponds to the lifetime measured after the samples are removed from the hot plate (90 μs before P-diffusion and 160 μs after P-diffusion). Subsequently, the samples are illuminated at 25°C for 40 hours, which already results in a slight decrease of τ to 64 μs and 135 μs, respectively (see Sec. 6.2.1). Afterwards, the lifetime is measured alternately after annealing in darkness at 200°C for time t (open symbols, corresponding to the annihilated state discussed in Chapter 5) and after illumination at 25°C for 40 hours (closed symbols, corresponding to the degraded state discussed in Chapter 4).

Figure 6.18 depicts a continuous destabilization of the deactivated state due to prolonged annealing in darkness: the longer the samples are annealed at 200°C in darkness, the lower the lifetime in the degraded state, suggesting that more and more 'permanently deactivated' BO centers are reactivated. Interestingly, we also observe a slight decrease of the lifetime measured directly after annealing (i.e. after full defect annihilation according to Chapter 5). This effect could be related to a slight decrease of the passivation quality of the silicon nitride layer, however, it could also indicate an increase in background defects.

6.3 Boron- and phosphorus-doped *n*-type Cz-Si

In Section 4.2.2, it was shown that light-induced degradation of the lifetime (at room temperature) is also observed in n-type Cz-Si doped with both boron and phosphorus. In Section 5.2.2, it was shown that this degradation is fully reversible by annealing in darkness. It can thus be concluded from our studies that boron-oxygen-related recombination centers also form in compensated n-type Cz-Si. However, it was also already shown in Sections 4.2.2 and 5.2.2 that the kinetics of both the defect generation and the defect annihilation in compensated n-type Cz-Si differ considerably from that in p-type Cz-Si (exclusively B-doped and compensated alike).

In this Section, the deactivation procedure is applied to compensated n-type Cz-Si for the first time. A set of six B- and P-doped n-type Cz-Si samples, all cut from the same ingot (Ingot B), is used. For sample details see Sec. 4.2.2 and Tab. 3.4. Processing steps include P-diffusion, removal of the resulting n^+ layers from both sides of the wafer, and surface passivation by PECVD-SiN$_x$. Figure 6.19(a) depicts the lifetime τ of three compensated n-type samples as a function of time. Photoconductance decay lifetime measurements are done using a Sinton Instruments WCT-120 setup and the lifetime is extracted at a fixed injection level of $\Delta p = 0.1 \times n_0$. During the first 10 hours, the samples are illuminated at 200°C at a light intensity of 100 mW/cm^2. Subsequently (as indicated by the dashed line), the samples are removed from the hot plate and illuminated at 25°C and 10 mW/cm^2.

Before the samples are placed under illumination on the hot plate, they are annealed at 200°C in darkness for 15 hours, to fully annihilate the boron-oxygen defect (see Sec. 5.2.2). During

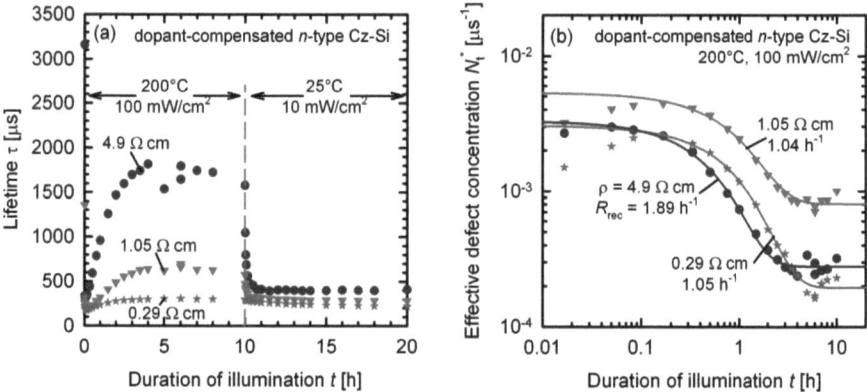

Figure 6.19: (a) Lifetime τ plotted versus time t of three B- and P-doped n-type Cz-Si samples with different resistivities. At first, the samples are illuminated for 10 hours at 200°C and at a light intensity of 100 mW/cm². Subsequently, indicated by the dashed red line, the samples are illuminated at 25°C and 10 mW/cm². (b) Effective defect concentration N_t^* during lifetime recovery, derived from the data shown in Fig. 6.19(a), plotted versus time t on a double logarithmic scale. The recovery rate constant of the lifetime R_{rec} is determined through fitting an exponential function to the data (solid lines).

illumination at 200°C, the lifetime first decreases, due to the accelerated generation of BO complexes, and subsequently increases again, similar to the evolution observed in p-type Cz-Si. After 10 h, the lifetimes saturate and the samples are removed from the hot plate (as indicated by the dashed red line). Surprisingly, the lifetimes are then unstable under illumination at room temperature, decreasing significantly during the following illumination at 25°C and 10 mW/cm².

Figure 6.19(b) depicts the effective defect concentration N_t^* derived from the lifetime data in Fig. 6.19(a) (up to the dashed red line). Fitting an exponential decay function of the form $y = y_0 + a \exp\left(-R_{rec} t\right)$ to the data, the recovery rate constant R_{rec} is obtained. In Fig. 6.20, the recovery rate constants R_{rec} of all samples are depicted as a function of the net doping concentration n_0 (red triangles). As can be seen from this plot, R_{rec} is comparable for all samples, even as n_0 varies over one order of magnitude.

Given that the recovered lifetime is not stable under illumination at room temperature, it might be possible that the observed recovery at 200°C under illumination is actually the same process as the defect annihilation during annealing in darkness. Figure 6.20 thus also includes the annihilation rate constants R_{ann} determined in the same samples during defect annihilation at 200°C in darkness (black circles, data taken from Fig. 5.7). At high net doping concentrations $n_0 \geq 2 \times 10^{16}$ cm^{-3} the annihilation rate constant R_{ann} and the recovery rate

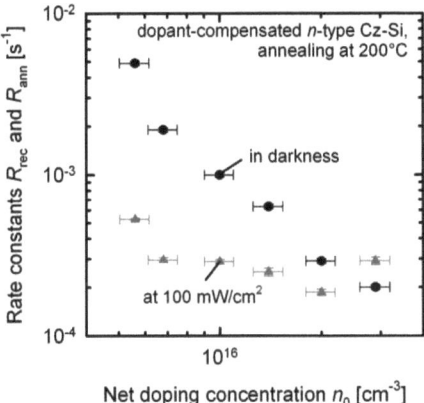

Figure 6.20: Recovery rate constants R_{rec} and annihilation rate constants R_{ann} in dopant-compensated n-type Cz-Si. Recovery was done at 200°C and a light intensity of 100 mW/cm^2 (red triangles) while defect annihilation was performed at 200°C in darkness (black circles).

constant R_{rec} are indeed very similar. At lower n_0, however, R_{ann} increases with decreasing n_0, while R_{rec} remains constant.

In addition, it should be noted that the evolution of the effective defect concentration N_t^* during annealing in darkness could not be described by a simple exponential decay, but had to be fitted using a stretched exponential function, where the stretching factor β varied between 0.58 and 0.92 (see Sec. 5.2.2). In conclusion, this comparison shows that the lifetime recovery at elevated temperature in darkness and the lifetime recovery at elevated temperature under illumination are not identical in compensated n-type Cz-Si.

6.4 Chapter summary

In this Chapter, the impact of processing steps and material parameters on the deactivation of the boron-oxygen-related recombination center was studied for the first time. A phosphorus diffusion was found to speed up the deactivation rate constant R_{de} by up to a factor of 4 and simultaneously decrease the defect concentration N_t^* after deactivation, thus resulting in higher lifetimes.

In addition, the impact of different plasma processes was investigated. The comparison between deposition of a PECVD silicon nitride (SiN$_x$) layer with and without exposure to the plasma during deposition revealed that the deactivation rate constant R_{de} in the samples with plasma exposure was one order of magnitude larger than in the samples without plasma exposure. On the other hand, no significant difference in R_{de} was observed between samples covered with PECVD-SiN$_x$ without plasma exposure and samples covered with plasma-assisted ALD-Al$_2$O$_3$. This suggests that the hydrogen content in the dielectric layer does not play a crucial role in the deactivation process by itself. Instead, a precursor for the deactivation process may be created in the presence of a plasma. Exposure to the plasma during deposition could then either increase the amount of this precursor or facilitate the in-diffusion of the precursor into the silicon bulk. A possible candidate for this precursor is some form of hydrogen, since hydrogen is present both during PA-ALD of Al$_2$O$_3$ and PECVD of SiN$_x$.

Concerning material parameters, it was found in this work that the deactivation rate constant R_{de} in exclusively boron-doped Cz-Si decreases with increasing boron concentration N_A as well as with increasing interstitial oxygen concentration [O$_i$]. The creation of thermal donors noticeably reduced the speed of the deactivation process and increased the defect concentration N_t^* after permanent deactivation. These results are important for identifying the defect reactions leading to the permanent deactivation of the boron-oxygen defect, as will be discussed in more detail in Chapter 7.

Regarding the carrier lifetime after permanent deactivation τ_{de}, a decrease of τ_{de} with increasing doping concentration N_A was observed. However, τ_{de} was also found to vary strongly, e.g. by up to a factor of 2.5 in samples that were cut from the same wafer and underwent the same processing steps. A fit to the highest lifetimes measured after permanent deactivation in this work yielded a power law $\tau_{\text{de}} = 10^{26} \, N_A^{-1.46}$ (τ_{de} in μs and N_A in cm^{-3}) for B-only doped Cz-Si samples that underwent a P-diffusion (and have an interstitial oxygen concentration of [O$_i$] = $(7.5 \pm 0.5) \times 10^{17}$ cm^{-3}). This dependence describes the potential of the carrier lifetime after permanent deactivation (τ_{de}) as a function of boron concentration N_A and will be used in Chapter 8 to estimate the efficiency potential of screen-printed solar cells fabricated on B-doped Cz-Si.

Finally, it was found that the permanent deactivation of the BO defect is not possible in compensated n-type Cz-Si doped with both boron and phosphorus.

7 Defect models

7.1 The B_sO_{2i}-model

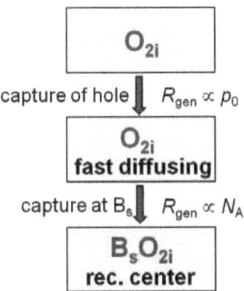

Figure 7.1: Schematic of the defect generation mechanism in the B_sO_{2i} defect model proposed in [19].

In non-compensated boron-doped Cz-Si, the defect concentration N_t^* shows a proportional increase with the total boron concentration N_A and a quadratic increase with the interstitial oxygen concentration $[O_i]$ [5, 14, 17, 18]. In addition, the defect generation rate constant R_{gen} was found to be proportional to N_A^2 [8, 9, 18]. These results led to the development of a defect model where the recombination center is composed of one substitutional boron atom B_s and an interstitial oxygen dimer O_{2i} [14, 15, 19]. A simplified schematic of the process of defect formation is depicted in Fig. 7.1.

The defect concentration N_t^* is naturally proportional to N_A in this model. Additionally, the defect kinetics of the B_sO_{2i}-model predict a proportionality of R_{gen} on the product of the hole concentration and the total boron concentration $p_0 \times N_A$. The dependence on p_0 follows from the requirement of the oxygen dimer to catch a hole to speed up its diffusivity. The configuration-coordinate diagram of the oxygen dimer diffusion is shown in Fig. 7.2.

The oxygen dimer exists in two configurations: squared (O_{2i}^{sq}) and staggered (O_{2i}^{st}) [15]. The diffusion is proposed to proceed via a Bourgoin mechanism [56] by alternately catching electrons and holes, which results in considerably reduced energy barriers for the conversion from one configuration to the other. In p-type silicon, the oxygen dimer is double-positively charged (blue line) and the energy barrier for the transition from the squared to the staggered configuration

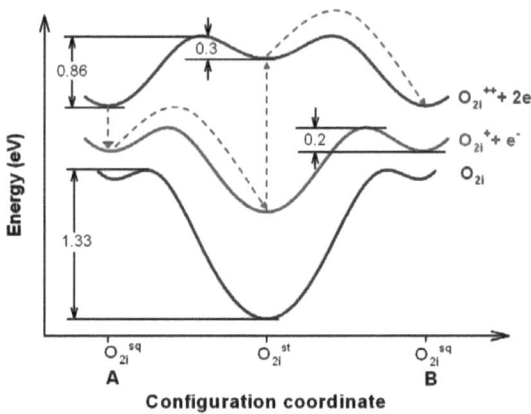

Figure 7.2: Configuration-coordinate diagram of the oxygen dimer diffusion [15].

is relatively high (0.86 eV), which would inhibit diffusion at room temperature.

However, under illumination, excess carriers are generated and the double-positive dimer catches an electron. In the single-positive charge state, the barrier from the squared to the staggered configuration significantly reduces to 0.2 eV (green line). After the reconfiguration, the dimer is then proposed to emit an electron, thus recharging to its double-positive state. From there, the energy barrier to the squared state is 0.3 eV. Through the alternate capture and emission of an electron, the dimer-diffusion energy is thus reduced to 0.3 eV, making dimer diffusion at room temperature possible.

The dependence of the generation rate constant on N_A is a consequence of the oxygen dimers bonding with a B_s atom [19]. This model was found to be in excellent agreement with the experimental data measured on non-compensated p-type Cz-Si, since in exclusively B-doped Cz-Si $p_0 \wedge N_A - N_A^2$.

However, in Chapter 4 new experimental data obtained on compensated silicon was presented, which revealed that N_t^* actually depends on p_0 (and not on N_A) and that R_{gen} actually depends on p_0^2 and not on $p_0 \times N_A$.

With regard to the defect annihilation in darkness, the B_sO_{2i}-model suggests a dissociation of the substitutional boron atom B_s and the interstitial oxygen dimer O_{2i}. However, in Chapter 5, we presented new experimental data on the annihilation process, which was obtained on compensated p-type Cz-Si, and which revealed that the defect annihilation rate constant is inversely proportional to the hole concentration p_0. This dependence cannot be explained with a dissociation process, in particular since it was proposed that the dissociation always takes place in the positive charge state of B_sO_{2i} [15].

The new experimental findings presented in Chapter 4 and 5 thus call for a reassessment of the B_sO_{2i}-model.

7.2 The B_iO_{2i}-model

Defect generation under illumination

Based on the new experimental findings presented in this thesis, an alternative defect model was recently proposed by Voronkov and Falster [57]. In this model, the recombination center is comprised of one interstitial boron atom B_i and an interstitial oxygen dimer O_{2i}. The interstitial boron concentration $[B_i]$, and consequently the defect concentration $[B_iO_{2i}]$, in this model is proposed to be proportional to the net doping concentration p_0, which is in agreement with the existing data on exclusively boron-doped Cz-Si as well as with our new data obtained on compensated p-type Cz-Si.

The dependence of $[B_i]$ on p_0 follows from the generation mechanism of interstitial boron atoms: during ingot cooling, B_i are created via the kick-out mechanism during the growth of oxygen precipitates. Subsequently, the interstitial boron atoms congregate to large clusters. However, at moderate temperatures, B_i^+ atoms (B_i is positively charged in p-type silicon) can also be released from the neutral clusters after capturing a hole h^+. As a result, there is a constant exchange between the two states [57]:

$$B_i^+ \, (dissolved) \rightleftharpoons B_i \, (cluster) + h^+ \tag{7.1}$$

Consequently, the concentration of (dissolved) interstitial boron atoms $[B_i^+]$ is proportional to the hole concentration $[h^+]$, which in turn equals the net doping concentration p_0 after the ingot temperature passes the intrinsic point. At lower temperature, the exchange between dissolved B_i^+ and B_i clusters eventually ceases and $[B_i^+]$ becomes fixed. The highly mobile interstitial boron atoms B_i^+ may then bind with other species such as O_i, O_{2i} and B_s. However, at room temperature, these complexes are all frozen in and the existing B_iO_{2i} complexes are the so-called latent form (LC) of the boron-oxygen defect.

Apart from explaining the dependence of the final defect concentration N_t^* on p_0, the model also proposes an explanation for the quadratic dependence of the defect generation rate constant R_{gen} on the net doping concentration p_0 [57]. The latent form of the complex (LC) exists in two charge states: single positive (LC^+) and neutral (LC^0). The energy minimum in darkness is at LC^+, which is thus the stable ground state. Under illumination, or rather in the presence of excess carriers, however, LC^+ is recharged to its neutral state LC^0 through the capture of an electron. In the neutral state, the free energy F of the latent center and the slow-forming recombination center (SRC) increase and the minimum is now at SRC^0.

However, in order to make the transition from LC to SRC, an energy barrier needs to be overcome, as indicated by the arrow in the schematic of the free energy profiles of LC and SRC shown in Fig. 7.3(a). The experimental finding that the defect generation rate constant R_{gen}

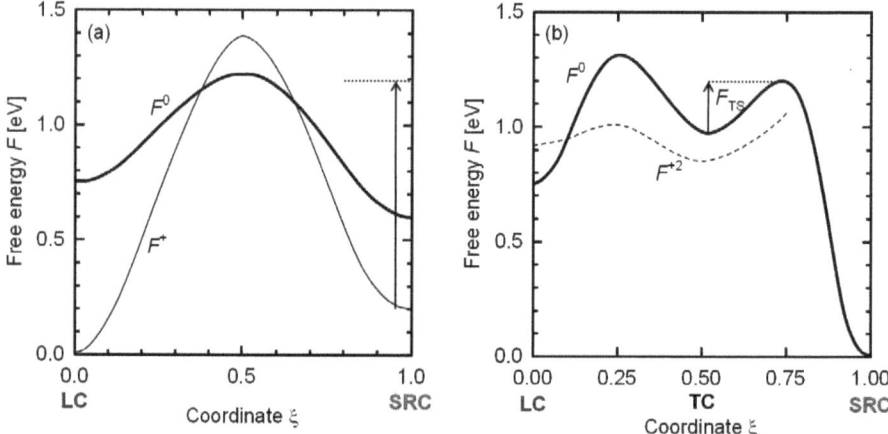

Figure 7.3: Free energy diagrams of the latent (LC) and the recombination-active (SRC) B_iO_{2i} defect [57].

is proportional to the square of the doping concentration p_0^2, suggests that the transformation from LC to SRC involves the capture of two holes. It can thus be assumed that the transition from LC to SRC proceeds over a transient center (TC), as depicted in Fig. 7.3(b).

The neutral LC^0 captures two holes h^+ and recharges to a double positive state LC^{+2}. In this state, the energy barrier to the transient center TC^{+2} is considerably reduced. TC^{+2} then captures two electrons and becomes TC^0. Through overcoming the barrier F_{ts}, TC^0 eventually transforms to SRC^0, i.e. the recombination-active form of the B_iO_{2i} defect. A simplified schematic of the process of defect formation is depicted in Fig. 7.4.

Note that the proposed B_iO_{2i}-model addresses the slow-forming recombination center (SRC). The fast forming recombination center (FRC), on the other hand, could be yet another configuration of B_iO_{2i} or a different grown-in defect, such as B_sO_{2i}. To clarify this point, the concentration of FRC should be investigated as a function of the boron concentration N_A and the doping concentration p_0. If FRC is another configuration of B_iO_{2i}, then its concentration would also depend on p_0. However, if FRC originates from B_sO_{2i}, then its concentration would depend on N_A. In addition, the dependence of the fast-forming defect generation rate constant on the boron and doping concentration should be examined.

Defect annihilation at elevated temperature in darkness

During defect annihilation at elevated temperature in darkness, the recombination-active form of B_iO_{2i} (SRC) reconstructs back into its latent form (LC), since in darkness, the defect is single-positive (SRC^+) and the energy minimum in that charge state lies at LC^+ [see Fig.

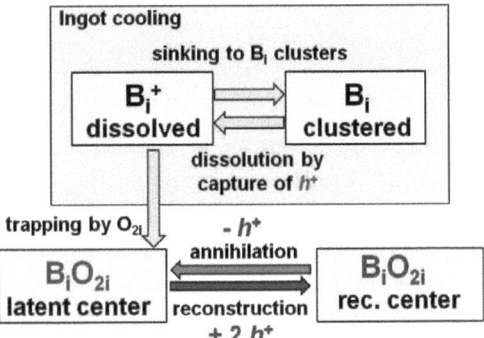

Figure 7.4: Schematic of the defect generation mechanism in the B_iO_{2i} defect model proposed in Ref. [57].

7.3(a)] [57]. Through thermal activation, SRC$^+$ moves along the free energy profile F^+ to the intersection point with the profile F^0. At this point, SRC$^+$ emits a hole and becomes SRC^0, which considerably reduces the energy barrier to LC. This small barrier is then overcome by thermal activation. At the second intersection of F^0 and F^+, the defect recharges back to the single-positive state and subsequently proceeds to its energy minimum at LC$^+$.

Due to the emission of a hole, the equilibrium concentration of neutral B_iO_{2i} is proportional to $1/p_0$ and the rate constant of the annihilation is expected to follow this dependence. This is in excellent agreement with the new experimental data presented in Chapter 5, where the investigation of compensated p-type Cz-Si revealed an inverse dependence of the defect annihilation rate constant on the net doping concentration p_0.

Permanent deactivation of the defect under illumination at elevated temperature

Permanent deactivation of the boron-oxygen-related recombination center under illumination at elevated temperature is attributed to the dissociation of the B_iO_{2i} defect and subsequent trapping of the interstitial boron atoms into B_i-clusters [57]. As was shown in Chapter 6, illumination at elevated temperature initially leads to a decrease of the lifetime (if the sample is not completely degraded already) due to the generation of boron-oxygen defects. Only after this initial decrease the lifetime increases again and is subsequently stable under illumination at room temperature.

This finding is in accord with the energy diagram depicted in Fig. 7.3(a). Under illumination, B_iO_{2i} is neutral and the energy minimum in that charge state lies at the recombination-active form (SRC). Accordingly, existing LC are expected to transform into SRC. Since illumination continues throughout the treatment, the B_iO_{2i} defects remain in the neutral charge state, where the energetically favorable configuration is SRC. As a result, the thermal activation (due to the elevated temperature) does not lead to a reconstruction to LC. Instead, the B_iO_{2i} slowly begin to dissociate. The dissolved B_i are highly mobile and may then either migrate to the B_i-clusters or be trapped into B_iO or B_iB_s. Through this, the overall concentration of B_iO_{2i} is permanently reduced.

Note that the observed recovery rate constants suggest that permanent recovery is controlled by sinking to the B_i-clusters [57]. This is in accordance with the new experimental findings presented in this thesis, since the deactivation rate constant was found to *decrease* with increasing boron concentration (see Sec. 6.1.3) and with increasing interstitial oxygen concentration (see Sec. 6.1.4). If the interstitial boron atoms were trapped by oxygen atoms O, the deactivation rate constant would be expected to increase with increasing $[O_i]$, and similarly with increasing $[B_s]$ in the case of trapping by B_s.

In addition, the acceleration of the deactivation process after a phosphorus diffusion (see Sec. 6.1.1) indicates an increase in a component which is involved in the deactivation process and the concentration of B_i-clusters could indeed by increased by a high temperature treatment such as a P-diffusion. On the other hand, it is conceivable that prolonged annealing at an intermediate temperature (e.g. 400-500°C) reduces the concentration of B_i-clusters through Ostwald ripening, i.e., through the dissolution of small B_i-clusters and subsequent sinking of the dissolved B_i to larger B_i-clusters. This mechanism would account for the observed reduction of the deactivation rate constant after thermal donor creation, as described in Section 6.1.5. Note that in this case, the reduction would actually be due to the reduction of B_i-clusters instead of the presence of thermal donors.

Likewise, the decrease of the deactivation rate constant with increasing doping concentration may be related to a lower concentration of B_i-clusters of greater size. Note, however, that the properties of the B_i-clusters are most likely governed either by the interstitial boron concentration $[B_i]$ or the hole concentration p_0 (or both) and not by the substitutional boron concentration $[B_s]$. While the results obtained on compensated p-type Cz-Si (Sec. 6.1.6) were ambiguous in that regard, an actual dependence of the deactivation rate constant on the hole concentration is certainly conceivable from the experimental data.

In summary, the new experimental findings presented in this thesis are consistent with the model of a slow dissociation of the B_iO_{2i} defects and subsequent migration of the dissolved B_i to B_i-clusters.

7.3 Chapter summary

In this Chapter, two defect models for the boron-oxygen-related recombination center were discussed. The first model, in which the recombination center is comprised of a substitutional boron atom B_s and an interstitial oxygen dimer O_{2i}, was found to be unable to explain the new experimental results obtained on compensated p-type Cz-Si in this work. The defect concentration N_t^* in this model is proportional to the substitutional boron concentration $[B_s]$, whereas the studies on compensated p-type silicon presented in Chapter 4 revealed that N_t^* is in fact proportional to the hole concentration p_0. In addition, the model proposes that defect generation proceeds via enhanced diffusion of the oxygen dimer under illumination and subsequent capture by a substitutional boron atom. As a result, the B_sO_{2i}-model predicts that the defect generation rate constant R_{gen} is proportional to the product of the boron concentration and the hole concentration, i.e. $R_{gen} \propto N_A \times p_0$. However, our studies on compensated p-type Cz-Si showed that R_{gen} is actually proportional to the square of the hole concentration p_0^2. Finally, the defect annihilation rate constant R_{ann} was found to be inversely proportional to the hole concentration, which is not consistent with the dissociation of the boron-oxygen defect proposed in the B_sO_{2i}-model.

Instead, a defect model in which the complex is comprised of an interstitial boron atom B_i and an interstitial oxygen dimer O_{2i} was found to be capable of explaining both the existing experimental results on non-compensated B-doped Cz-Si and the new results obtained on compensated p-type Cz-Si. In this model, the defect concentration N_t^* is proportional to the interstitial boron concentration $[B_i]$, which in turn is proportional to the hole concentration p_0. Defect generation in this model proceeds via a reconstruction of the B_iO_{2i} defect from the so-called latent form to the recombination-active form. This reconstruction involves the capture of two holes, thus resulting in a quadratic dependence of the defect generation rate constant R_{gen} on the hole concentration p_0. In turn, defect annihilation in darkness is a reconstruction back into the latent form of the defect. The inverse dependence of the defect annihilation rate constant on the hole concentration is attributed to the emission of a hole during the reconstruction. Accordingly, the B_iO_{2i}-model agrees with the new experimental data presented in this thesis.

8 Application of the deactivation procedure to solar cells

In order to study the permanent deactivation of the boron-oxygen complex in solar cells, the deactivation process is applied to screen-printed solar cells, which constitute the majority of crystalline silicon solar cells fabricated today. Using lifetime data obtained in Chapter 6, the potential of the permanent deactivation for such solar cells is evaluated by one-dimensional solar cell simulations using PC1D. In addition, the deactivation procedure is also applied to high-efficiency RISE-EWT solar cells, which subsequently achieve stable efficiencies above 20% on low-resistivity B-doped Czochralski-grown silicon.

8.1 Screen-printed solar cells

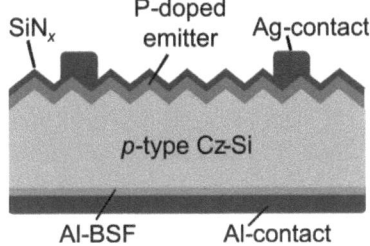

Figure 8.1: Schematic of a simple screen-printed solar cell with a full-area back surface field.

Figure 8.1 depicts a schematic of a simple screen-printed solar cell. The base material is boron-doped p-type Cz-Si. The phosphorus-doped emitter on the front-side of the cell is contacted via silver fingers, while the rear of the cell is fully covered with aluminum in order to contact the base. Fabrication steps include RCA cleaning, front-surface texturing, phosphorus diffusion and anti-reflective coating with silicon nitride (SiN_x), as well as screen-printing and co-firing of the contacts. Additionally, during the firing step, an Al-doped p^+ region is formed at the rear of the cell; the so-called back surface field (BSF).

In order to reduce the harmful impact of the boron-oxygen-related recombination center, today's production uses p-type Cz-Si with a resistivity of about 3 Ω cm. Using the simple cell design shown in Fig. 8.1, efficiencies of up to 18.7% are achievable [58].

8.1.1 Exclusively boron-doped Cz-Si

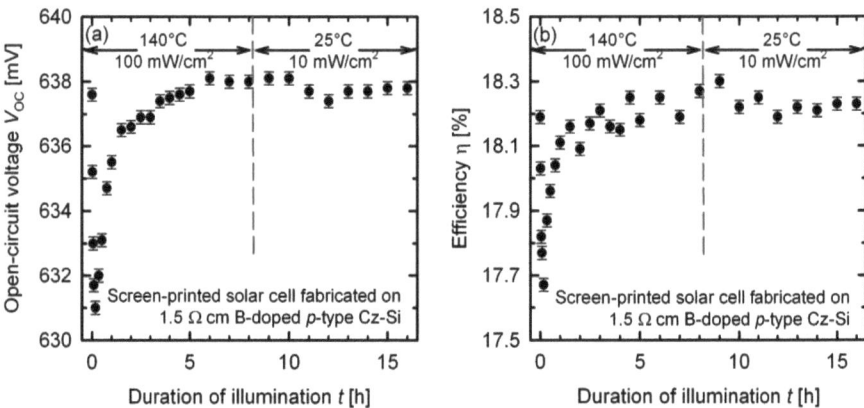

Figure 8.2: Time dependence of (a) the open-circuit voltage V_{OC}, and (b) the efficiency η of a screen-printed solar cell with a full-area BSF made on 1.5 Ω cm B-doped p-type Cz-Si under illumination at 140°C and a light intensity of 100 mW/cm². After removal from the hot plate (indicated by the dashed line), no efficiency degradation is discernible after several hours of illumination at room temperature and 10 mW/cm² light intensity.

To study the BO-deactivation in solar cells made on exclusively boron-doped Cz-Si, a screen-printed solar cell fabricated on 1.5 Ω cm boron-doped p-type Cz-Si is examined. The solar cell has an area of 125 × 125 mm² and is characterized by current-voltage measurements using a pv-tools LOANA system. IV-characteristics are measured under illumination of an LED array (wavelength 950 nm) at 25°C. Note that even though the illumination is done at a single wavelength, the measurement setup is very close to standard testing conditions (AM1.5 spectrum at 100 mW/cm² light intensity) as the intensity of the LED array is set using a calibrated solar cell with similar spectral response.

The calibrated solar cell is illuminated by a flash with a spectrum comparable to the AM1.5 spectrum. During the flash, the short-circuit current of the calibrated solar cell is measured as a function of light intensity. Using the known short-circuit current under 100 mW/cm² of the AM1.5 spectrum, the light intensity of the flash that corresponds to 100 mW/cm² AM1.5 is determined. Subsequently, the solar cell of interest is flashed and the short-circuit current is measured as a function of light intensity. Using the previously determined light intensity of the flash that corresponds to 100 mW/cm² AM1.5, the short-circuit current of the investigated solar cell under 100 mW/cm² AM1.5 is obtained. This value is then used to adjust the light intensity of the LED array.

Before light-induced degradation (i.e. after a 5 minute anneal at 200°C in darkness), the solar cell has an open-circuit voltage V_{OC} of (637.1 ± 0.2) mV and an efficiency η of $(18.19 \pm 0.02)\,\%$. After complete degradation, the open-circuit voltage is (630.3 ± 0.2) mV and the efficiency is $(17.67 \pm 0.02)\,\%$.

Deactivation is done at 140°C and an illumination intensity of 100 mW/cm². The evolution of both the open-circuit voltage V_{OC} and the efficiency η are shown in Fig. 8.2(a) and (b), respectively. Similar to the lifetime τ in lifetime samples, the open-circuit voltage initially decreases under illumination at 140°C due to the accelerated formation of the BO defect. After 10 minutes, however, V_{OC} and η begin to increase.

After 8 hours, the open-circuit voltage V_{OC} saturates at (638.0 ± 0.2) mV, while the efficiency η saturates at $(18.27 \pm 0.02)\,\%$. Note that the open-circuit voltage and the efficiency after deactivation are thus higher than after defect annihilation at 200°C in darkness. This deviation is most likely a result of the fast stage of degradation [12, 17], which cannot be avoided during solar cell measurements due to the need for residual light during contacting of the solar cell. Consequently, the values of V_{OC} and η which are attributed to the annihilated state may be reduced by the effects of the fast-forming defect.

In order to test the stability of the deactivated state, the solar cell is subsequently illuminated at 25°C and a light intensity of 10 mW/cm². As can be seen in Fig. 8.2, no degradation is observed over the course of 8 hours.

8.1.2 Boron- and phosphorus-doped Cz-Si

The material used to study permanent deactivation in solar cells made on compensated p-type Cz-Si is 1.45 Ω cm B- and P-doped Cz-Si from Ingot A. The wafers were cut from 10% to 16% relative distance from the seed end and were subsequently subjected to an industrial solar cell process. The resulting screen-printed solar cells have an area of 156 × 156 mm² and are characterized by current-voltage measurements using the pv-tools LOANA system.

Before light-induced degradation, the energy conversion efficiency η of all solar cells is $(17.1 \pm 0.1)\,\%$, while the open-circuit voltage V_{OC} is (614.7 ± 0.6) mV. After complete degradation $\eta = (15.8 \pm 0.2)\,\%$ and $V_{OC} = (601.5 \pm 1.2)$ mV. Before deactivation, all cells are annealed for 5 minutes at 200°C in darkness, thus annihilating all boron-oxygen defects.

Figure 8.3 depicts (a) the open-circuit voltage V_{OC} and (b) the efficiency η as a function of time t during illumination. During the first four hours (as indicated by the dashed line), the solar cell is illuminated at 140°C and a light intensity of 100 mW/cm². Subsequently, the solar cell is illuminated at 25°C and a light intensity of 10 mW/cm² in order to test the stability of the deactivated state.

During the first few minutes, both the open-circuit voltage V_{OC} and the efficiency η decrease, due to the formation of BO complexes. After 20 minutes, however, V_{OC} and η start to increase until they begin to saturate after four hours. Subsequent illumination at 25°C results in no significant decrease of either V_{OC} or η, indicating that no formation of BO complexes occurs.

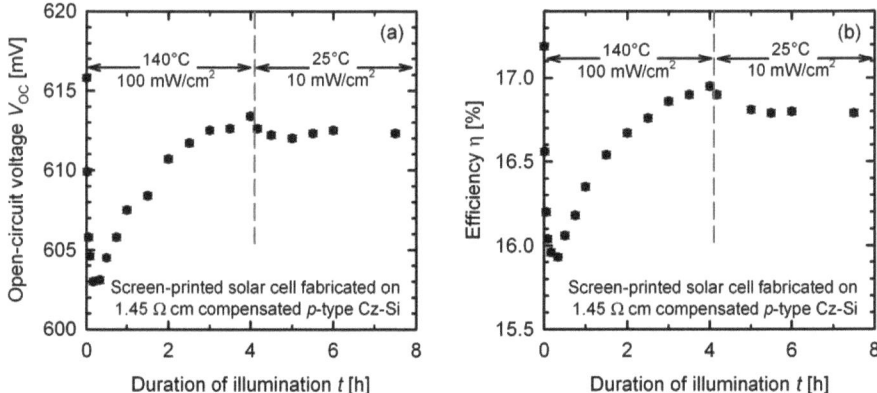

Figure 8.3: Time dependence of (a) the open-circuit voltage V_{OC}, and (b) the efficiency η of a screen-printed solar cell with a full-area BSF made on 1.45 Ω cm dopant-compensated p-type Cz-Si under illumination at 100 mW/cm^2 at 140°C. After removal from the hot plate (indicated by the dashed line), no efficiency degradation is discernible after several hours of illumination at room temperature.

Defect deactivation is also done at 165°C, 185°C and 200°C, respectively. The resulting deactivation rate constants R_{de} are plotted versus the inverse temperature $1000/T$ in Fig. 8.4. The data shown in Fig. 8.4 is fitted by an Arrhenius law

$$R_{de} = \nu \exp\left(-E_{de}/k_B T\right), \tag{8.1}$$

where ν is the attempt frequency and E_{de} is the activation energy of the deactivation process. From the fit shown in Fig. 8.4 an activation energy of $E_{de} = (0.60 \pm 0.11)$ eV is obtained. This is comparable to the activation energy determined in Section 6.1.1 on non-compensated lifetime samples of similar resistivity.

In order to evaluate the impact of the solar cell process on the deactivation process, Fig. 8.5 depicts a comparison of R_{de} data obtained on lifetime samples made on the same material as the investigated solar cells and of the data from Fig. 8.4. As was already shown in Section 6.1.1, a phosphorus diffusion increases the speed at which the deactivation proceeds by up to a factor of 4. This increase can also be observed in the present case, when comparing R_{de} data obtained on as-grown samples (black circles) and on samples that underwent a P-diffusion (red triangles).

Looking at the deactivation rate constants R_{de} measured on solar cells (blue triangles down), no additional increase of R_{de} due to the additional processing steps (such as firing of the contacts) are observed. However, at lower temperatures (140°C and 165°C), the deactivation rate constants are similar to those measured on P-diffused lifetime samples (red triangles),

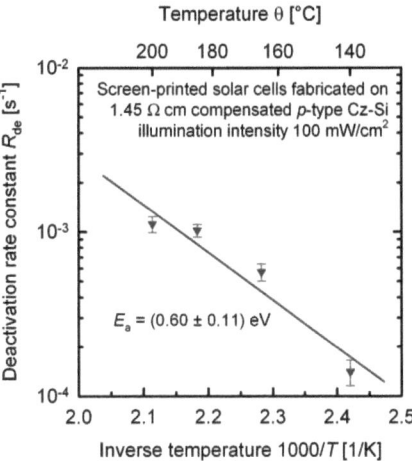

Figure 8.4: Arrhenius plot of the deactivation rate constants R_{de} determined on screen-printed solar cells fabricated on 1.45 Ω cm dopant-compensated p-type Cz-Si. An activation energy of $E_{de} = (0.60 \pm 0.11)$ eV is obtained.

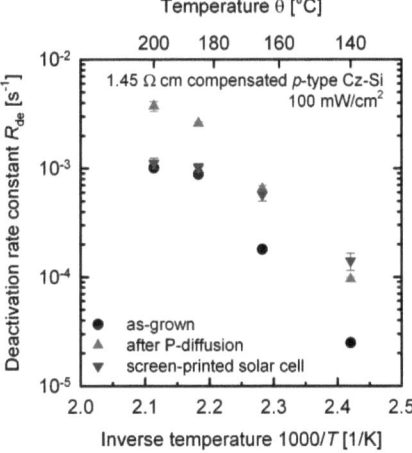

Figure 8.5: Comparison of deactivation rate constants R_{de} determined on lifetime samples before (black circles) and after P-diffusion (red triangles up) as well as on screen-printed solar cells (blue triangles down) fabricated on 1.45 Ω cm dopant-compensated p-type Cz-Si. While the P-diffusion increases R_{de} by a factor of 4, additional processing steps during solar cell production do not seem to have an impact on the deactivation rate constant.

whereas at higher temperatures (185°C and 200°C), the deactivation rate constants R_{de} of the solar cells are similar to those of the as-grown samples (black circles).

8.2 Efficiency potential of screen-printed solar cells according to one dimensional solar cell simulations

In the previous Section, it was shown that screen-printed solar cells with a full-area BSF which are fabricated on low-resistivity B-doped p-type Cz-Si suffer from a loss in energy conversion efficiency η of 0.5 to more than 1% absolute. In contrast, industrial screen-printed solar cells fabricated on high-resistivity (around 3 Ω cm) B-doped p-type Cz-Si typically degrade by only 0.1 to 0.2% absolute. As a result, the efficiency after light-induced degradation is higher in the high-resistivity solar cells than in the low-resistivity solar cells.

However, after permanent deactivation of the boron-oxygen defect low-resistivity material is expected to yield higher energy conversion efficiencies than high-resistivity material, especially when using advanced solar cell concepts such as the passivated emitter and rear cell (PERC) structure [59]. A schematic of the PERC concept is shown in Fig. 8.6. One of the main loss mechanisms in a simple screen-printed solar cell with a full-area BSF (as shown in Fig. 8.1) is recombination at the rear, because the back surface field provides only weak passivation. In the PERC concept, on the other hand, the full area back surface field is replaced by a full area passivating dielectric (e.g. a SiO_2/SiN_x stack), by which the recombination losses are considerably reduced. However, as the majority-carriers still need to be extracted from the base, the dielectric layer is locally removed (e.g. by laser ablation) in order to make room for the contacts.

In order to minimize the recombination losses at the rear, the contact area should obviously be as small as possible. On the other hand, small contact areas result in longer paths for the majority-carriers to the contacts. Too high resistivities of the base material thus result in series resistance losses. As a result, locally contacted solar cells require low-resistivity material for optimal performance. However, since the concentration of boron-oxygen-related recombination centers is proportional to the doping concentration, PERC fabricated on B-doped p-type Cz-Si suffer from pronounced light-induced degradation, making them the ideal candidate for application of the permanent deactivation treatment discussed in Chapter 6.

The potential of screen-printed PERC after application of the permanent deactivation is estimated by one-dimensional device simulations using PC1D. The input parameters for the simulations are derived from the characterization of a real passivated emitter and rear solar cell, which has an energy conversion efficiency of 19.4% [60]. The thickness of the solar cell is 180 μm. The front of the solar cell is textured with random pyramids and the n^+ emitter at the front has a sheet resistance of 80 $\Omega/\square$ with a peak concentration at the surface of 1×10^{20} cm^{-3}. The emitter is passivated by a SiO_2/SiN_x stack. Accordingly, the surface recombination

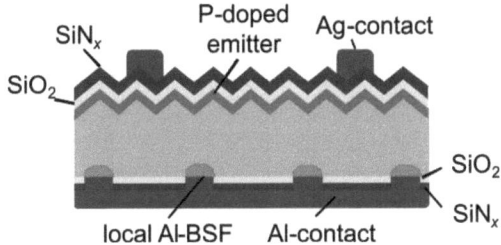

Figure 8.6: Schematic of the passivated emitter and rear cell (PERC) concept.

velocity at the front S_{front} is estimated to be 10^4 cm/s.

At the rear, a similar passivation stack of SiO_2/SiN_x is used. To allow contacting, the passivation is locally removed by laser ablation, resulting in 130 μm wide openings with a pitch (i.e. distance between the lines) of 2 mm. Subsequently, silver contacts are screen-printed to the front while aluminum is screen-printed to the back. During the firing step, local back surface fields form in the local contact openings. The resulting effective surface recombination velocity at the rear S_{rear} is determined to be 70 cm/s [60].

Note that the optical model implemented in the PC1D program is inadequate. Consequently, the measured total reflectance of the solar cell is used to take the optical parameters of the cell into account. In addition, the 19.4% solar cell suffers from a low fill factor FF of 75.8%. The loss in fill factor is due to an increased specific series resistance of 1.6 Ω cm² [58], the bigger part of which can be attributed to an increased specific contact resistance to the base.

By varying the metallization fraction, a specific contact resistance to the base of 40 to 50 mΩ cm² was recently found for local screen-printed contacts [58]. In contrast, specific contact resistances of 10 mΩ cm² are measured for full-area screen-printed aluminum contacts. Assuming that the specific contact resistance for local contacts can be optimized to 20 mΩ cm², we set the specific contact resistance to the base at 0.3 Ω cm² for the PC1D simulation (taking into account the metallization fraction of 0.065). The specific series resistance arising from resistance losses in the emitter, contact resistance to the emitter, and series resistance of the finger grid and busbars is set at 0.4 Ω cm². A list of the used simulation parameters is summarized in Tab. 8.1.

The impact of the base resistivity on the specific series resistance is calculated according to Plagwitz [61]. Taking into account the wafer thickness $W = 180\mu$m, the contact line width and the pitch, the specific series resistance originating from the base doping R_{base} is obtained as a function of the doping concentration p_0, as depicted in Fig. 8.7. At low doping concentrations p_0, the specific series resistance R_{base} drastically increases, up to a value of 2.25 Ω cm² at a doping concentration of $p_0 = 1 \times 10^{15}$ cm^{-3}. In current production, screen-printed solar cells are fabricated on 3 Ω cm B-doped p-type Cz-Si, in order to avoid the detrimental impact of lifetime degradation. This corresponds to a doping concentration of $p_0 = 4.7 \times 10^{15}$ cm^{-3} and

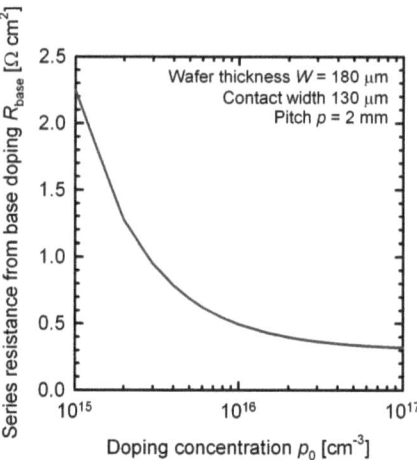

Figure 8.7: Specific series resistance R_base resulting from the base doping concentration p_0 in a locally contacted solar cell with a thickness W of 180 μm, contact width of 130 μm and a pitch p (i.e., distance between the contact lines) of 2 mm, calculated according to Plagwitz [61].

a resulting specific series resistance of 0.70 Ω cm², which would lead to significant losses in the fill factor of the solar cell.

To illustrate the benefit of the deactivation treatment for screen-printed passivated emitter and rear solar cells, simulations are done for carrier lifetimes in the degraded state as well as for carrier lifetimes after permanent deactivation of the boron-oxygen defect. The carrier lifetime after permanent deactivation of the BO defect at a given resistivity is taken from the assessment of the lifetime potential presented in Section 6.1.7, while the lifetime in the degraded state is calculated according to Bothe et al. [17].

Note that in the degraded state, the bulk carrier lifetime shows a strong dependence on the excess carrier density, due to the asymmetrical capture cross sections for electrons and holes ($\sigma_n/\sigma_p = 10$) [6, 13]. In the PC1D program, this dependence is taken into account through the capture time constants for electrons τ_{n0} and holes τ_{p0}, as well as through the energy level of the defect E_t. According to Bothe, the capture time constant for electrons after a phosphorus diffusion step is given by $\tau_{n0} = 1.48 \cdot 7.675 \times 10^{45} \, N_\text{A}^{-0.824} \, [O_\text{i}]^{-1.748}$ (τ_{n0} in μs and N_A and $[O_\text{i}]$ in cm^{-3}), whereas the capture time constant for holes is $\tau_{p0} = 10 \, \tau_{n0}$ [23]. In addition, the energy level of the BO defect lies at $E_\text{t} = E_\text{C} - 0.41$ eV [13].

After permanent deactivation of the boron-oxygen defect, the dependence of the bulk carrier lifetime on the excess carrier density is much weaker, suggesting that the carrier lifetime in that state is limited by a defect with similar capture cross sections for electrons and holes. As a consequence, $\tau_{n0} = \tau_{p0} = 10^{26} \, N_\text{A}^{-1.46}$ (τ in μs and N_A in cm^{-3}) is assumed for the PC1D

Table 8.1: Input parameters for one dimensional solar cell simulations of a passivated emitter and rear cell using PC1D.

General:	Base thickness: 180 μm
	Specific emitter contact resistance: 0.40 Ω cm^2
Optics:	taken into account by measured total reflectance
Bulk:	recombination in degraded state [23]:
	$\tau_{n0} = 1.48 \cdot 7.675 \times 10^{45} N_A^{-0.824} (7.5 \times 10^{17})^{-1.748}$
	$\tau_{p0} = 10\, \tau_{n0}$
	$E_t - E_i = 0.15$ eV
	recombination after permanent deactivation:
	$\tau_{n0} = \tau_{p0} = 10^{26}\, N_A^{-1.46}$
	$E_t - E_i = 0$
Emitter:	80 $\Omega/\square$; error function profile
	peak concentration 1×10^{20} cm^{-3}; depth factor: 0.16 μm
Surfaces:	$S_\text{front} = 10^4$ cm/s; $S_\text{rear} = 70$ cm/s

simulation of the deactivated state.

The results of the solar cell simulations using PC1D are depicted in Fig. 8.8. The short-circuit current density J_SC notably decreases with increasing doping concentration, both in the degraded state (black solid line) and after permanent deactivation of the BO defect (red dashed line), as shown in Fig. 8.8(a). However, J_SC values above 38.5 mA/cm^2 are possible up to a doping concentration of $p_0 = 2 \times 10^{16}$ cm^{-3} after permanent deactivation of the BO defect due to the increased lifetimes. In the degraded state, on the other hand, J_SC values above 38.5 mA/cm^2 are only possible with doping concentrations below 2.3×10^{15} cm^{-3}.

In Fig. 8.8(b), the open-circuit voltage V_OC is depicted, which has a more complex dependence on the doping concentration p_0. In general, V_OC increases with increasing base doping concentration p_0. However, this effect is counteracted by a decrease of V_OC due to the decreasing bulk carrier lifetime. As a result, the open-circuit voltage in the degraded state has a minimum of $V_\text{OC} = 629$ mV around a doping concentration of $p_0 = 10^{16}$ cm^{-3}. After application of the permanent deactivation the impact of the bulk lifetime is considerably reduced, resulting in a steady increase of V_OC with increasing p_0 up to a doping concentration of $p_0 = 2 \times 10^{16}$ cm^{-3}, where $V_\text{OC} = 668$ mV is achieved. At even higher doping concentrations, however, the reduction in bulk carrier lifetime again becomes noticeable, resulting in a decrease of V_OC.

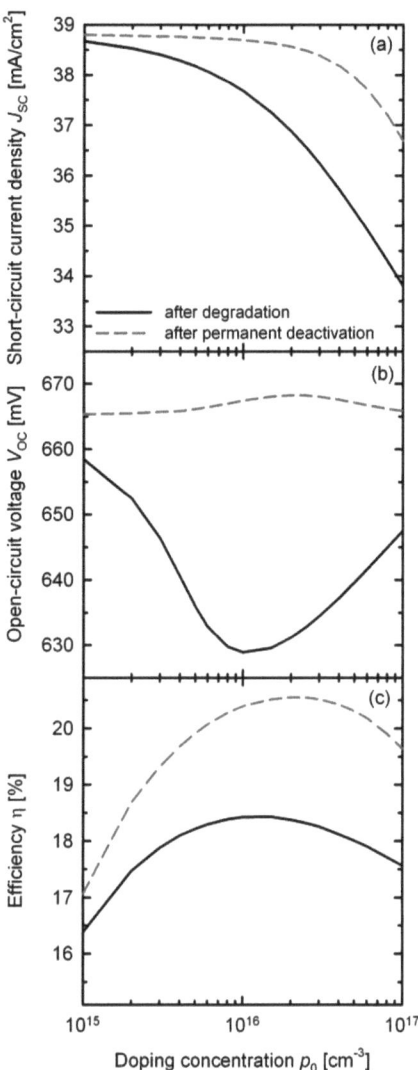

Figure 8.8: Simulated solar cell parameters of a screen-printed passivated emitter and rear cell (PERC) as a function of doping concentration p_0 after complete degradation (solid black lines) and after permanent deactivation of the boron-oxygen defect (dashed red lines), respectively. (a) Short-circuit current density J_{SC}, (b) open-circuit voltage V_{OC}, and (c) energy conversion efficiency η.

Looking only at the short-circuit current density J_{SC} and the open-circuit voltage V_{OC}, one would expect the maximum efficiency in the degraded state at low doping concentrations. However, due to the drastically increased series resistance at high base resistivity, the losses in fill factor completely diminish the high J_{SC} and V_{OC} values. As a result, the highest simulated efficiency of 18.4% in the degraded state is obtained for a doping concentration of $p_0 = 1.5 \times 10^{16}$ cm^{-3}, as shown in Fig. 8.8(c).

After permanent deactivation, the improved short-circuit current densities at moderate doping concentrations shift the maximum in energy conversion efficiency to an even higher doping concentration of $p_0 = 2 \times 10^{16}$ cm^{-3}, where an efficiency of 20.6% is simulated.

Through application of the permanent deactivation process, it is thus possible to achieve energy conversion efficiencies of up to 20.6% on a screen-printed PERC structure using 0.8 Ω cm B-doped p-type Cz-Si.

8.3 High-efficiency RISE-EWT solar cells

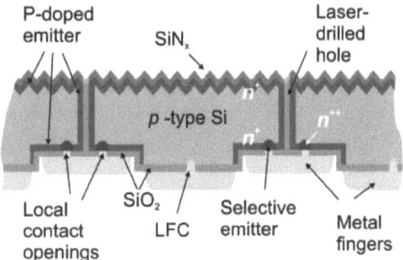

Figure 8.9: Schematic of a RISE-EWT solar cell [62]

In the last Section, an efficiency potential of 20.6% was simulated for screen-printed solar cells using the passivated emitter and rear cell (PERC) concept. In order to obtain efficiencies above 21%, however, more sophisticated solar cell designs are needed. One of these high-efficiency concepts is the Rear Interdigitated Single Evaporation Emitter Wrap-Through (RISE-EWT) cell [63]. A schematic of the RISE-EWT cell structure is shown in Fig. 8.9. Base and emitter fingers are interdigitated and on different height levels on the rear side of the cell. The n^+ emitter on the front is connected to the emitter on the back via laser-drilled holes (Emitter Wrap-Through structure [64]). The base region is contacted by laser-fired contacts (LFC) [65], which are formed through laser-firing of aluminum through a silicon oxide passivation layer.

Fabrication steps include RCA cleaning, oxidation, front-surface texturing, laser structuring of the rear, laser drilling of holes, phosphorus diffusion, front-surface passivation, a single metallization step of the entire rear with self-aligning contact separation [63], and formation of the LFCs. Details of the solar cell process can be found in Ref. [62].

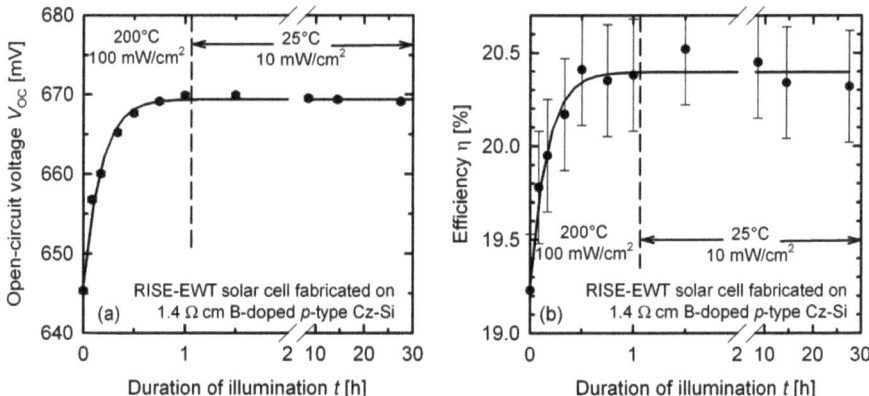

Figure 8.10: Time dependence of (a) the open-circuit voltage V_{OC}, and (b) the efficiency η of a high-efficiency RISE-EWT solar cell made on 1.4 Ω cm B-doped Cz-Si under illumination at 100 mW/cm^2 and 200°C. After removal from the hot plate (indicated by the dashed line), no efficiency degradation is discernible after several hours of illumination at room temperature.

The base material of the cells investigated in this study is 1.4 Ω cm boron-doped p-type Cz-Si. Current-voltage characteristics are measured under standard testing conditions (25°C, AM1.5 spectrum, 100 mW/cm^2), which are set using a calibrated RISE-EWT solar cell with a similar spectral response. The total cell area is 100 cm^2, however, due to a non-optimized busbar design the measurements presented here refer to a designated area of 92 cm^2. Measurements performed on the total cell area yield the same open-circuit voltage and short-circuit current density while the fill factor suffers a loss of 2% absolute.

RISE-EWT solar cells made on FZ-Si have demonstrated energy conversion efficiencies of up to 21.4% on a designated area of 92 cm^2 [62]. On low-resistivity (1-2 Ω cm) boron-doped Cz-Si, efficiencies of (20.0 ± 0.5) % are typically achieved before light-induced degradation and (19.0 ± 0.5) % are measured after complete degradation on the same illuminated cell area.

Figure 8.10 shows the evolution of (a) the open-circuit voltage V_{OC} and (b) the efficiency η of an exemplary RISE-EWT cell. The initial efficiency before light-induced degradation was (20.2 ± 0.3)%. After 8 h of illumination at room temperature using a halogen lamp at 20 mW/cm^2, this value decreased to (19.2 ± 0.3)%. The open-circuit voltage after degradation is (645.3 ± 0.3) mV and the short-circuit current density 40.4 mA/cm^2. Note that the short-circuit current density shows only a negligible degradation because for EWT-type cells the short-circuit current density is less sensitive to reduced base diffusion lengths, and thus to reduced bulk lifetimes, than for standard cells.

In order to deactivate the boron-oxygen-related recombination center, the solar cell is illuminated at 200°C with a halogen lamp at a light intensity of 100 mW/cm^2. As can be seen in

Fig. 8.10, V_{OC} increases to (669.1 ± 0.3) mV after 60 min. J_{SC} remains the same within the uncertainty range (40.9 mA/cm^2), resulting in an energy conversion efficiency of $(20.4 \pm 0.3)\%$. After twenty hours of illumination at room temperature the IV-characteristic is taken again, showing no noticeable loss in either V_{OC} or η. No further efficiency loss due to the formation of boron-oxygen complexes is thus expected. Through application of the deactivation treatment to RISE-EWT solar cells, a stable efficiency above 20% was therefore achieved for the first time on low-resistivity B-doped p-type Cz-Si.

8.4 Chapter summary

In this Chapter, the deactivation procedure was applied to solar cells. The deactivation rate constants measured on screen-printed solar cells fabricated on exclusively B-doped p-type Cz-Si as well as on dopant-compensated p-type Cz-Si were similar to those obtained on lifetime samples that underwent a P-diffusion. Taking into account the lifetime potential found in Chapter 6 and using realistic solar cell parameters obtained through characterization of a real screen-printed passivated emitter and rear cell (PERC), one-dimensional device simulations were performed to estimate the efficiency potential of screen-printed PERC fabricated on B-doped p-type Cz-Si. Through this, a maximum efficiency of 20.6% was simulated for 0.8 Ω cm B-doped p-type base material. In addition, the application of the deactivation procedure to high-efficiency RISE-EWT solar cells resulted in the first stable efficiencies above 20% on 1.4 Ω cm B-doped p-type Cz-Si.

9 Summary

In this work, crystalline silicon doped with boron and phosphorus was investigated. The carrier mobility in such compensated silicon was found to be significantly reduced when compared to non-compensated crystalline silicon with similar resistivity. Besides an overall reduction of 25% (in monocrystalline Czochralski-grown silicon) and 45% (in block-cast multicrystalline silicon), an additional significant decrease of the carrier mobilities was observed at very high compensation ratios (i.e. at very similar boron and phosphorus concentrations). The overall reduction was attributed to the increased amounts of ionized scattering centers, while the additional decrease was attributed to reduced screening of these scattering centers at low free carrier concentrations.

Using the quasi-steady-state photoconductance technique, the carrier lifetime was studied under illumination at room temperature, during annealing in darkness, and under illumination at elevated temperature. Illumination at room temperature has long been known to result in a degradation of the carrier lifetime in exclusively boron-doped oxygen-rich silicon, such as B-doped Czochralski-grown silicon, and has been studied in boron and phosphorus-doped Czochralski silicon for the first time in this thesis. Since the doping concentration in compensated silicon follows from the difference of the boron and phosphorus concentrations, the impact of the boron and the doping concentration can be investigated separately in compensated silicon.

Through this, it was revealed in this work that the effective defect concentration is actually proportional to the doping concentration and not, as previously assumed, to the boron concentration. In addition, the defect generation rate constant was found to depend quadratically on the doping concentration. Since these findings could not be explained by the standard defect model for the boron-oxygen-related recombination center, the standard defect model needed to be reassessed.

Light-induced degradation of the carrier lifetime was also studied for the first time in compensated n-type silicon, i.e. silicon that contains more phosphorus than boron, in this thesis. The time scale of the defect generation was found to be 50 h to 100 h. Interestingly, both the speed of the degradation and the effective defect concentration after degradation were found to be independent of the net doping concentration. In addition, the time dependence of the effective defect concentration was found to be quite complex and could not be described by a simple exponential function. However, such a complex behavior can be attributed to a dependence of the generation rate constant on the hole concentration. Since holes are minority carriers in n-type silicon, their concentration depends on the minority carrier lifetime, which in turn depends on the concentration of recombination-active defects.

Annealing in darkness at elevated temperature results in a recovery of the degraded lifetime. However, this state is unstable under illumination. In studies on exclusively boron-doped silicon, the rate constant of this annihilation process was found to decrease with increasing boron concentration. In this work, studies on compensated p-type silicon revealed that the dependence is actually on the doping concentration. Additionally, defect annihilation in compensated n-type silicon was found to take considerably longer than in p-type silicon, with an increase in annihilation rate constant by up to a factor of 1000.

Illumination at elevated temperature results in a permanent deactivation of the boron-oxygen center. In exclusively boron-doped silicon, the deactivation rate constant was found to depend inversely on the boron concentration and inversely on the square of the interstitial oxygen concentration. In addition, a phosphorus diffusion was found to increase the deactivation rate constant by up to a factor of 4. Deposition of a silicon nitride layer by plasma-enhanced chemical vapor deposition increased the deactivation rate constant by a factor of 5, when the samples were exposed to the plasma during deposition.

Permanent deactivation of the boron-oxygen-related recombination center in compensated n-type silicon was not observed. While the carrier lifetime in compensated n-type silicon does increase under illumination at elevated temperature, subsequent illumination at room temperature results in renewed degradation of the lifetime. Importantly, partial and even complete degradation of the carrier lifetime after application of the deactivation treatment was also observed in p-type silicon. Prolonged illumination at elevated temperature as well as extended annealing in darkness were found to destabilize the deactivated state, resulting in renewed degradation of the lifetime under illumination at room temperature.

The new experimental results obtained in this thesis were discussed with regard to the standard defect model for the boron-oxygen-related recombination center, in which the defect is composed of one substitutional boron atom B_s and an interstitial oxygen dimer O_{2i}. Since this model predicts a linear dependence of the defect concentration on the boron concentration, this model was incapable of explaining the new results obtained on compensated silicon in this work. In addition, the B_sO_{2i}-model predicts that the defect generation rate constant is proportional to the product of the boron and the hole concentration, whereas the new studies on compensated p-type silicon revealed a quadratic dependence of the generation rate constant on the hole concentration. Instead, an alternative defect model, in which the defect is composed of one interstitial boron atom B_i and an interstitial oxygen dimer O_{2i}, was found to be in agreement with the new experimental data.

Finally, the permanent deactivation treatment was applied to industrial screen-printed solar cells as well as high-efficiency Rear Interdigitated Single Evaporation Emitter Wrap-Through

(RISE-EWT) solar cells fabricated on 1.4 to 1.5 Ω cm p-type Cz-Si. A complete recovery of the open-circuit voltage and the solar cell efficiency were observed on both exclusively B-doped silicon and on dopant-compensated silicon. As a result, an increase of the solar cell efficiency by 0.5% absolute was observed on screen-printed solar cells fabricated on 1.5 Ω cm exclusively boron-doped Cz-Si, while and increase of 1.3% absolute was achieved on industrial screen-printed solar cells made on 1.45 Ω cm compensated p-type Cz-Si. In addition, one-dimensional device simulations showed an efficiency potential of 20.6% for screen-printed passivated emitter and rear cells fabricated on 0.8 Ω cm B-doped p-type Cz-Si after permanent deactivation of the BO defect. Using the high-efficiency RISE-EWT solar cell concept, a stabilized efficiency of 20.4% was achieved on 1.4 Ω cm B-doped p-type Cz-Si after application of the permanent deactivation treatment.

References

[1] *PHOTON* **4**, 38 (2010).

[2] H. Fischer and W. Pschunder, Investigation of photon and thermal induced changes in silicon solar cells, *Proc. 10th IEEE PVSC* p. 404 (1973).

[3] J. Knobloch, S. W. Glunz, V. Henninger, W. Warta, W. Wettling, F. Schomann, W. Schmidt, A. Endrös, and K. A. Münzer, 21% efficient solar cells processed from Czochralski grown silicon, *Proc. 13th EUPVSEC* p. 9 (1995).

[4] J. Schmidt, A. G. Aberle, and R. Hezel, Investigation of carrier lifetime instabilities in Cz-grown silicon, *Proc. 26th IEEE PVSC* p. 13 (1997).

[5] S. W. Glunz, S. Rein, W. Warta, J. Knobloch, and W. Wettling, On the degradation of Cz-silicon solar cells, *Proc. 2nd WCPEC* p. 1343 (1998).

[6] J. Schmidt and A. Cuevas, Electronic properties of light-induced recombination centers in boron-doped Czochralski silicon, *J. Appl. Phys.* **86**, 3175–3180 (1999).

[7] S. Glunz, S. Rein, and J. Knobloch, Stable Czochralski-grown silicon solar cells using gallium-doped base materials, *Proc. 16th EUPVSEC* p. 1070 (2000).

[8] S. W. Glunz, S. Rein, J. Y. Lee, and W. Warta, On the degradation of Cz-silicon solar cells, *J. Appl. Phys.* **90**, 2397 (2001).

[9] S. Rein, T. Rehrl, W. Warta, S. W. Glunz, and G. Willeke, Electrical and thermal properties of the metastable defect in boron-doped Czochralski silicon (Cz-Si), *Proc. 17th EUPVSEC* p. 1555 (2001).

[10] J. Schmidt, K. Bothe, and R. Hezel, Formation and annihilation of the metastable defect in boron-doped Czochralski silicon, *Proc. 29th IEEE PVSC* p. 178 (2002).

[11] K. Bothe, J. Schmidt, and R. Hezel, Effective reduction of the metastable defect concentration in boron-doped Czochralski silicon for solar cells, *Proc. 29th IEEE PVSC* p. 194 (2002).

[12] H. Hashigami and T. Saitoh, Carrier-induced degradation phenomena of carrier lifetime and cell performance in boron-doped Cz-Si, *Proc. 3rd WCPEC* S5O–C6–02 (2003).

[13] S. Rein and S. W. Glunz, Electronic properties of the metastable defect in boron-doped Czochralski silicon: Unambiguous determination by advanced lifetime spectroscopy, *Appl. Phys. Lett.* **82**, 1054 (2003).

[14] J. Schmidt and K. Bothe, Structure and transformation of the metastable boron- and oxygen-related defect center in crystalline silicon, *Phys. Rev. B* **69**, 024107 (2004).

[15] J. Adey, R. Jones, D. W. Palmer, P. R. Briddon, and S. Öberg, Degradation of boron-doped Czochralski-grown silicon solar cells, *Phys. Rev. Lett.* **93**, 055504 (2004).

[16] K. Bothe, R. Hezel, and J. Schmidt, Understanding and reducing the boron-oxygen-related performance degradation in Czochralski silicon solar cells, *Solid State Phenomena* **95-96**, 223 (2004).

[17] K. Bothe, R. Sinton, and J. Schmidt, Fundamental boron–oxygen-related carrier lifetime limit in mono- and multicrystalline silicon, *Prog. Photovolt: Res. Appl.* **13**, 287 (2005).

[18] K. Bothe and J. Schmidt, Electronically activated boron-oxygen-related recombination centers in crystalline silicon, *J. Appl. Phys.* **99**, 013701 (2006).

[19] D. W. Palmer, K. Bothe, and J. Schmidt, Kinetics of the electronically stimulated formation of a boron-oxygen complex in crystalline silicon, *Phys. Rev. B* **76**, 035210 (2007).

[20] A. Herguth, G. Schubert, M. Kaes, and G. Hahn, Avoiding boron-oxygen related degradation in highly boron doped Cz silicon, *Proc. 21st EUPVSEC* p. 530 (2006).

[21] A. Herguth, G. Schubert, M. Kaes, and G. Hahn, Investigations on the long time behavior of the metastable boron–oxygen complex in crystalline silicon, *Prog. Photovolt: Res. Appl.* **16**, 135 (2008).

[22] R. Sinton and A. Cuevas, Contactless determination of current–voltage characteristics and minority-carrier lifetimes in semiconductors from quasi-steady-state photoconductance data, *Appl. Phys. Lett.* **69**, 2510 (1996).

[23] K. Bothe, *Oxygen-related trapping and recombination centres in boron-doped crystalline silicon*, PhD thesis, University of Hanover (2006).

[24] P. P. Altermatt, J. Schmidt, M. Kerr, G. Heiser, and A. G. Aberle, Exciton-enhanced Auger recombination in crystalline silicon under intermediate and high injection conditions, *Proc. 16th EUPVSEC* p. 247 (2000).

[25] K. Peter, R. Kopecek, A. Soiland, and E. Enebakk, Future potential for SoG-Si feedstock from the metallurgical process route, *Proc. 23rd EUPVSEC* p. 947 (2008).

[26] J. Libal, S. Novaglia, M. Acciarri, S. Binetti, R. Petres, J. Arumughan, R. Kopecek, and A. Prokopenko, Effect of compensation and of metallic impurities on the electrical properties of Cz-grown solar grade silicon, *J. Appl. Phys.* **104**, 104507 (2008).

[27] J. Veirman, S. Dubois, N. Enjalbert, J. P. Garandet, D. R. Heslinga, and M. Lemiti, Hall mobility reduction in single-crystalline silicon gradually compensated by thermal donors activation, *Solid-State Electron.* **54**, 671–674 (2010).

[28] H. Nagel, C. Berge, and A. G. Aberle, Generalized analysis of quasi-steady-state and quasi-transient measurements of carrier lifetimes in semiconductors, *J. Appl. Phys.* **86**, 6218 (1999).

[29] C. Berge, Vergleich transienter und quasistatischer Photoleitfähigkeitsmessungen an kristallinem Silicium, Master's thesis, University of Hanover, (1998).

[30] K. R. McIntosh and R. Sinton, Uncertainty in photoconductance lifetime measurements that use an inductive coil detector, *Proc. 23rd EUPVSEC* p. 77 (2008).

[31] S. M. Sze, *Physics of semiconductor devices*, John Wiley & Sons (1981).

[32] A. B. Sproul, M. A. Green, and A. W. Stephens, Accurate determination of minority carrier- and lattice scattering-mobility in silicon from photoconductance decay, *J. Appl. Phys.* **72**, 4161 (1992).

[33] E. Peiner, A. Schlachetzki, and D. Krüger, Doping profile analysis in Si by electrochemical capacitance-voltage measurements, *J. Electrochem. Soc.* **142**, 576 (1995).

[34] R. Bock, P. P. Altermatt, and J. Schmidt, Accurate extraction of doping profiles from electrochemical capacitance voltage measurements, *Proc. 23rd EUPVSEC* p. 1510 (2008).

[35] J. Czochralski, Ein neues Verfahren zur Messung der Kristallisationsgeschwindigkeit der Metalle, *Z. Phys. Chem.* **92**, 219 (1918).

[36] A. A. Istratov, T. Buonassisi, R. J. McDonald, A. R. Smith, R. Schindler, J. A. Rand, J. P. Kalejs, and E. R. Weber, Metal content of multicrystalline silicon for solar cells and its impact on minority carrier diffusion length, *J. Appl. Phys.* **94**, 6552 (2003).

[37] D. Macdonald, A. Cuevas, A. Kinomura, Y. Nakano, and L. J. Geerligs, Transition-metal profiles in a multicrystalline silicon ingot, *J. Appl. Phys.* **97**, 033523 (2005).

[38] E. Scheil, Bemerkungen zur Schichtkristallbildung, *Z. Metallk.* **34**, 70 (1942).

[39] D. Long and J. Myers, Ionized-impurity scattering mobility of electrons in silicon, *Phys. Rev.* **115**, 1107 (1959).

[40] D. B. M. Klaassen, A unified mobility model for device simulation - I. Model equations and concentration dependence, *Sol. Stat. Electronics* **35**, 953 (1992).

[41] D. B. M. Klaassen, A unified mobility model for device simulation - II. Temperature dependence of carrier mobility and lifetime, *Sol. Stat. Electronics* **35**, 961 (1992).

[42] D. Macdonald, A. Cuevas, and L. J. Geerligs, Measuring dopant concentrations in compensated p-type crystalline silicon via iron-acceptor pairing, *Appl. Phys. Lett.* **92**, 202119 (2008).

[43] F. Shimura, *Semiconductor Silicon Crystal Technology*, Academic Press (1980).

[44] T. Lauinger, *Untersuchung und Optimierung neuartiger Plasmaverfahren zur Siliciumnitrid-Beschichtung von Silicium-Solarzellen*, PhD thesis, University of Hanover (2001).

[45] B. Lenkeit, *Elektronische und strukturelle Eigenschaften von Plasma-Siliziumnitrid zur Oberflächenpassivierung von siebgedruckten, bifazialen Silizium-Solarzellen*, PhD thesis, University of Hanover (2002).

[46] K. A. Münzer, Hydrogenated silicon nitride for regeneration of light induced degradation, *Proc. 24th EUPVSEC* p. 1558 (2009).

[47] G. Dingemans, M. C. M. van de Sanden, and W. M. M. Kessels, Influence of the deposition temperature on the c-Si surface passivation by Al_2O_3 films synthesized by ALD and PECVD, *Electrochem. Sol. Lett.* **13**, H76 (2010).

[48] J. Schmidt, B. Veith, and R. Brendel, Effective surface passivation of crystalline silicon using ultrathin Al_2O_3 films and Al_2O_3/SiN_x stacks, *phys. stat. sol. (RRL)* **3**, 287 (2009).

[49] G. Agostinelli, A. Delabie, P. Vitanov, Z. Alexieva, H. F. W. Dekkers, S. De Wolf, and G. Beaucarne, Very low surface recombination velocities on p-type silicon wafers passivated with a dielectric with fixed negative charge, *Sol. Energy Mater. Sol. Cells* **90**, 3438 (2006).

[50] B. Hoex, S. B. S. Heil, E. Langereis, M. C. M. van de Sanden, and W. M. M. Kessels, Ultralow surface recombination of c-Si substrates passivated by plasma-assisted atomic layer deposited Al_2O_3, *Appl. Phys. Lett.* **89**, 042112 (2006).

[51] B. Sopori, M. I. Symko, R. Reedy, K. Jones, and R. Matson, Mechanism(s) of hydrogen diffusion in silicon solar cells during forming gas anneal, *Proc. 26th IEEE PVSC* p. 25 (1997).

[52] M. Stavola, F. Jiang, S. Kleekajai, L. Wen, C. Peng, V. Yelundur, A. Rohatgi, G. Hahn, L. Carnel, and J. Kalejs, Hydrogen passivation of defects in crystalline silicon solar cells, *Mater. Res. Soc. Symp. Proc.* **1210**, 1210–Q01–01 (2010).

[53] C. S. Fuller, J. A. Ditzenberger, N. B. Hannay, and E. Buehler, Resistivity changes in silicon single crystals induced by heat treatment, *Acta Metall.* **3**, 97 (1955).

[54] W. Kaiser, Electrical and optical properties of heat-treated silicon, *Phys. Rev.* **105**, 1751 (1957).

[55] A. Herguth and G. Hahn, Boron–oxygen related defects in Cz-Si solar cells: degradation, regeneration and beyond, *Proc. 24th EUPVSEC* p. 974 (2009).

[56] J. C. Bourgoin and J. W. Corbett, A new mechanism for interstitial migration, *Phys. Lett.* **38A**, 135 (1972).

[57] V. V. Voronkov and R. Falster, Latent complexes of interstitial boron and oxygen dimers as a reason for degradation of silicon-based solar cells, *J. Appl. Phys.* **107**, 053509 (2010).

[58] S. Gatz, T. Dullweber, and R. Brendel, Evaluation of series resistance losses in screen-printed solar cells with local rear contacts, *IEEE J-PV* **1**, 37 (2011).

[59] A. W. Blakers, A. Wang, A. M. Milne, J. Zhao, and M. Green, 22.8% efficient silicon solar cell, *Appl. Phys. Lett.* **55**, 1363 (1989).

[60] S. Gatz, H. Hannebauer, R. Hesse, F. Werner, A. Schmidt, T. Dullweber, J. Schmidt, K. Bothe, and R. Brendel, 19.4%-efficient large-area fully screen-printed silicon solar cells, *phys. stat. sol. (RRL)* **5**, 147 (2011).

[61] H. Plagwitz, *Surface passivation of crystalline silicon solar cells by amorphous silicon films*, PhD thesis, University of Hanover (2007).

[62] S. Hermann, P. Engelhart, A. Merkle, T. Neubert, T. Brendemühl, R. Meyer, N.-P. Harder, and R. Brendel, 21.4%-efficient emitter wrap-through RISE solar cell on large area and picosecond laser processing of local contact openings, *Proc. 22nd EUPVSEC* p. 970 (2007).

[63] P. Engelhart, A. Teppe, A. Merkle, R. Grischke, R. Meyer, N.-P. Harder, and R. Brendel, The RISE-EWT solar cell – A new approach towards simple high efficiency silicon solar cells, *Proc. 15th PVSEC* p. 802 (2005).

[64] J. M. Gee, W. K. Schubert, , and P. A. Basore, Emitter wrap-through solar cell, *Proc. 23rd IEEE PVSC* p. 265 (1993).

[65] E. Schneiderlöchner, R. Preu, R. Lüdemann, S. W. Glunz, and G. Willeke, Laser-fired contacts (LFC), *Proc. 17th EUPVSEC* p. 1303 (2001).

List of Figures

2.1 Example of the injection-dependent carrier lifetime measured using the quasi-steady-state photoconductance technique. 9
2.2 Schematic of the setup used for quasi-steady-state photoconductance measurements. 10
2.3 Calibration function of the rf-bridge circuit. 11
2.4 Flow chart of the data analysis during quasi-steady-state photoconductance measurements. 12
2.5 Schematic of a microwave-detected photoconductance decay setup. 15

3.1 Schematic of the Czochralski-growth process. 20
3.2 Schematic of the block-casting process. 21
3.3 Dopant profiles in a compensated ingot according to the Scheil equation. . . . 23
3.4 Majority-carrier mobility in a compensated mc-Si ingot plotted versus the ingot height. 25
3.5 Effective carrier lifetime mapping of an as-cut wafer cut vertically from a compensated mc-Si ingot. 26
3.6 Minority-carrier mobility mapping of a wafer cut vertically from a compensated mc-Si ingot. 27
3.7 Hole mobility in a compensated Cz-Si ingot plotted versus the ingot height. . . . 29
3.8 Electron mobility in a compensated Cz-Si ingot plotted versus the ingot height. 30

4.1 Light-induced degradation of the lifetime in a 1.4 Ω cm B-doped p-type Cz-Si sample. 37
4.2 Comparison of light-induced degradation in dopant-compensated p- and n-type Cz-Si. 41
4.3 Light-induced degradation of the lifetime in dopant-compensated p-type Cz-Si. . 43
4.4 Effective defect concentration after complete degradation in dopant-compensated p-type Cz-Si, plotted versus the net doping concentration and the boron concentration, respectively. 44
4.5 Defect generation rate constant in dopant-compensated p-type Cz-Si as a function of the square of the net doping concentration and the product of net doping and boron concentration, respectively. 45

4.6　Light-induced degradation of the lifetime in dopant-compensated n-type Cz-Si. . 46

5.1　Lifetime and effective defect concentration in B-doped p-type Cz-Si as a function of time during defect annihilation at 140°C in darkness. 50
5.2　Annihilation rate constants in exclusively B-doped p-type Cz-Si determined at 140°C in darkness as a function of doping concentration. 50
5.3　Comparison of the defect annihilation in dopant-compensated p- and n-type Cz-Si. 51
5.4　Lifetime and effective defect concentration as a function of time during defect annihilation in dopant-compensated p-type Cz-Si. 52
5.5　Annihilation rate constants in dopant-compensated p-type Cz-Si determined at 140°C in darkness as a function of net doping and boron concentration, respectively. 53
5.6　Lifetime and effective defect concentration as a function of time during defect annihilation in dopant-compensated n-type Cz-Si at 200°C in darkness. 54
5.7　Annihilation rate constants in dopant-compensated n-type Cz-Si determined at 200°C in darkness as a function of net doping concentration. 55
5.8　Stretching factor as a function of net doping concentration. 55

6.1　Lifetime in B-doped p-type Cz-Si under illumination at 25°C and 185°C, respectively. 58
6.2　Comparison of lifetime in B-doped p-type Cz-Si before and after P-diffusion during illumination at 135°C. 60
6.3　Arrhenius plot of the deactivation rate constant in 1.4 Ω cm B-doped p-type Cz-Si before and after a P-diffusion step. 61
6.4　Comparison between the effective defect concentration after deactivation in as-grown samples and samples that underwent phosphorus diffusion. 61
6.5　Comparison of the permanent deactivation of the BO-defect in samples passivated with PECVD-SiN$_x$ and ALD-Al$_2$O$_3$, respectively. 63
6.6　Arrhenius plot of the deactivation rate constants determined in samples passivated with PECVD-SiN$_x$ and ALD-Al$_2$O$_3$, respectively. 64
6.7　Lifetime and effective defect concentration in three exclusively B-doped p-type Cz-Si samples with different resistivities as a function of time during illumination at 185°C. 66
6.8　Deactivation rate constants in exclusively B-doped p-type Cz-Si at 185°C as a function of boron concentration. 67
6.9　Activation energy of the deactivation process in exclusively B-doped p-type Cz-Si as a function of boron concentration. 67
6.10　Lifetime and normalized defect concentration in 0.72 Ω cm B-doped p-type Cz-Si with varying interstitial oxygen concentrations as a function of time during illumination at 185°C. 69

6.11 Deactivation rate constants in B-doped p-type Cz-Si as a function of interstitial oxygen concentration. 69
6.12 Lifetime and effective defect concentration as a function of time during illumination at 185°C in B-doped p-type Cz-Si with varying thermal donor concentration. 71
6.13 Lifetime and effective defect concentration as a function of time during illumination at 140°C in dopant-compensated p-type Cz-Si. 72
6.14 Deactivation rate constants in dopant-compensated p-type Cz-Si determined at 140°C as a function of net doping and boron concentration, respectively. 73
6.15 Potential of the carrier lifetime after permanent deactivation of the BO defect as a function of boron concentration. 74
6.16 Partial degradation of the carrier lifetime in B-doped p-type Cz-Si under illumination at room temperature after application of the deactivation treatment. . . . 76
6.17 Complete degradation of the carrier lifetime in B-doped p-type Cz-Si during long-term illumination at 185°C. 77
6.18 Complete degradation of the carrier lifetime in B-doped p-type Cz-Si through extended annealing in darkness. 78
6.19 Lifetime and effective defect concentration as a function of time during illumination at 100 mW/cm^2 and 200°C in dopant-compensated n-type Cz-Si. 80
6.20 Recovery and annihilation rate constants of the lifetime in dopant-compensated n-type Cz-Si at 200°C as a function of net doping concentration. 81

7.1 Schematic of the defect generation mechanism in the B_sO_{2i} defect model. 83
7.2 Configuration-coordinate diagram of the oxygen dimer diffusion. 84
7.3 Free energy diagrams of the latent and the recombination-active B_iO_{2i}-defect. . . 86
7.4 Schematic of the defect generation mechanism in the B_iO_{2i} defect model. 87

8.1 Schematic of a simple screen-printed solar cell with a full-area BSF. 91
8.2 Open-circuit voltage and efficiency of a screen-printed solar cell made on 1.5 Ω cm B-doped p-type Cz-Si during illumination at 140°C. 92
8.3 Open-circuit voltage and efficiency of a screen-printed solar cell made on 1.45 Ω cm dopant-compensated p-type Cz-Si during illumination at 140°C. 94
8.4 Arrhenius plot of the deactivation rate constants R_{de} determined on screen-printed solar cells fabricated on 1.45 Ω cm dopant-compensated p-type Cz-Si. . . 95
8.5 Comparison of deactivation rate constants determined on lifetime samples and on screen-printed solar cells fabricated on 1.45 Ω cm dopant-compensated p-type Cz-Si at different temperatures. 95
8.6 Schematic of the passivated emitter and rear cell concept. 97
8.7 Specific series resistance resulting from the base doping concentration in locally contacted solar cell structures. 98

8.8 Simulated solar cell parameters of a screen-printed passivated emitter and rear cell after complete degradation and after permanent deactivation, respectively. . 100

8.9 Schematic of a RISE-EWT solar cell. 101

8.10 Open-circuit voltage and efficiency of a RISE-EWT solar cell during illumination at 200°C. 102

List of publications

Publications arising from the work in this thesis:

Refereed journal papers

1. B. Lim, K. Bothe, and J. Schmidt, Deactivation of the boron-oxygen recombination center in silicon by illumination at elevated temperature, *phys. stat. sol. RRL* **2**, 93 (2008)

2. B. Lim, S. Hermann, K. Bothe, J. Schmidt, and R. Brendel, Solar cells on low-resistivity boron-doped Czochralski-grown silicon with stabilized efficiencies of 20%, *Appl. Phys. Lett.* **93**, 162102 (2008)

3. D. Macdonald, F. Rougieux, A. Cuevas, B. Lim, J. Schmidt, M. Di Sabatino, and L. J. Geerligs, Light-induced boron-oxygen defect generation in compensated p-type Czochralski silicon, *J. Appl. Phys.* **105**, 093704 (2009)

4. B. Lim, A. Liu, D. Macdonald, K. Bothe, and J. Schmidt, Impact of dopant compensation on the deactivation of boron-oxygen recombination centers in crystalline silicon, *Appl. Phys. Lett.* **95**, 232109 (2009)

5. B. Lim, K. Bothe, and J. Schmidt, Impact of oxygen on the permanent deactivation of boron-oxygen-related recombination centers in crystalline silicon, *J. Appl. Phys.* **107**, 123707 (2010)

6. F. E. Rougieux, D. Macdonald, A. Cuevas, S. Ruffell, J. Schmidt, B. Lim, and A. P. Knights, Electron and hole mobility reduction and Hall factor in phosphorus-compensated p-type silicon, *J. Appl. Phys.* **108**, 013706 (2010)

7. B. Lim, F. Rougieux, D. Macdonald, K. Bothe, and J. Schmidt, Generation and annihilation of boron-oxygen-related recombination centers in compensated p- and n-type silicon, *J. Appl. Phys.* **108**, 103722 (2010)

8. B. Lim, M. Wolf, and J. Schmidt, Carrier mobilities in multicrystalline silicon wafers made from UMG-Si, *phys. stat. sol. (c)* **8**, 835 (2011)

9. D. Macdonald, A. Liu, F. E. Rougieux, A. Cuevas, B. Lim, and J. Schmidt, The impact of dopant compensation on the boron-oxygen defect in p- and n-type crystalline silicon, *phys. stat. sol. (a)* **208**, 559 (2011)

10. B. Lim, V. V. Voronkov, R. Falster, K. Bothe, and J. Schmidt, Lifetime recovery in p-type Czochralski silicon due to the reconfiguration of boron–oxygen complexes via a hole-emitting process, *Appl. Phys. Lett.* **98**, 162104 (2011)

11. V. V. Voronkov, R. Falster, K. Bothe, B. Lim, and J. Schmidt, Lifetime-degrading boron-oxygen centres in p-type and n-type compensated silicon, *J. Appl. Phys.* **110**, 063515 (2011)

12. B. Lim, K. Bothe, and J. Schmidt, Accelerated deactivation of the boron-oxygen-related recombination center in crystalline silicon, *Semi. Sci. Tech.* **26**, 095009 (2011)

13. F. E. Rougieux, B. Lim, J. Schmidt, M. Forster, D. Macdonald, and A. Cuevas, Influence of net doping, excess carrier density and annealing on the boron oxygen related defect density in compensated n-type silicon, *J. Appl. Phys.* **110**, 063708 (2011)

Refereed papers presented at international conferences

1. B. Lim, K. Bothe, and J. Schmidt, Modeling the generation and dissociation of the boron-oxygen complex in B-doped Cz-Si, *Proceedings of the 33rd IEEE Photovoltaic Specialists Conference*, 151 (2008)

2. B. Lim, S. Hermann, K. Bothe, J. Schmidt, and R. Brendel, Permanent deactivation of the boron-oxygen recombination center in silicon solar cells, *Proceedings of the 23rd European Photovoltaic Solar Energy Conference*, p. 1018 (2008)

3. D. Macdonald, A. Liu, F. Rougieux, A. Cuevas, B. Lim, J. Schmidt, M. Di Sabatino, and L. J. Geerligs, Boron-oxygen defects in compensated p-type Czochralski silicon, *Proceedings of the 24th European Photovoltaic Solar Energy Conference*, p. 877 (2009)

4. B. Lim, A. Liu, F. Rougieux, D. Macdonald, K. Bothe, and J. Schmidt, Generation and annihilation of boron-oxygen-related recombination centers in compensated p- and n-type silicon, *Proceedings of the 25th European Photovoltaic Solar Energy Conference*, p. 1205 (2010)

5. F. Rougieux, D. Macdonald, A. Cuevas, S. Ruffell, J. Schmidt, B. Lim, and A. Knights, Electron and hole mobility reduction and Hall factor in compensated p-type silicon, *Proceedings of the 25th European Photovoltaic Solar Energy Conference*, p. 1244 (2010)

Danksagung

Diese Arbeit habe ich während meiner Tätigkeit am Institut für Solarenergieforschung Hameln (ISFH) angefertigt. Ich bedanke mich daher bei Professor Dr. Brendel und vor allem bei Professor Dr. Schmidt für die sehr gute wissenschaftliche Betreuung in dieser Zeit. Im Laufe meiner Tätigkeit wurde ich außerdem von vielen Kolleginnen und Kollegen am ISFH unterstützt. Dafür bedanke ich mich an dieser Stelle noch einmal ganz herzlich!

Wichtige Teile dieser Arbeit sind außerdem in Zusammenarbeit mit der Gruppe "Semiconductor and Solar Cells" an der Australian National University in Canberra unter der Leitung von Professor Dr. Cuevas entstanden.

Besonderer Dank gilt darüber hinaus Dr. Karsten Bothe sowie Dr. Robert Falster und Dr. Vladimir Voronkov für viele rege Diskussionen und einen stetigen Ideenaustausch.

Zuletzt möchte ich mich bei Professor Dr. Pfnür für die Übernahme des Korreferats bedanken sowie bei Professor Dr. Lechtenfeld für die Übernahme des Prüfungsvorsitzes bei meiner Disputation.

i want morebooks!

Buy your books fast and straightforward online - at one of world's fastest growing online book stores! Environmentally sound due to Print-on-Demand technologies.

Buy your books online at
www.get-morebooks.com

Kaufen Sie Ihre Bücher schnell und unkompliziert online – auf einer der am schnellsten wachsenden Buchhandelsplattformen weltweit! Dank Print-On-Demand umwelt- und ressourcenschonend produziert.

Bücher schneller online kaufen
www.morebooks.de

VDM Verlagsservicegesellschaft mbH
Heinrich-Böcking-Str. 6-8   Telefon: +49 681 3720 174   info@vdm-vsg.de
D - 66121 Saarbrücken   Telefax: +49 681 3720 1749   www.vdm-vsg.de

Printed by Books on Demand GmbH, Norderstedt / Germany